Advances of Future IoE Wireless Network Technology

Advances of Future IoE Wireless Network Technology

Editors

Gwo-Jiun Horng
S.T. Aripriharta
Yao-Tung Tsou
Chia-Wei Tsai

Basel • Beijing • Wuhan • Barcelona • Belgrade • Novi Sad • Cluj • Manchester

Editors

Gwo-Jiun Horng
Department of Computer
Science and
Information Engineering,
Southern Taiwan University
of Science and Technology,
Yung Kung,
Tainan, Taiwan

S.T. Aripriharta
Department of
Electrical Engineering,
Universitas Negeri Malang,
Malang, Indonesia

Yao-Tung Tsou
Department of
Communications
Engineering,
Feng Chia University,
Taichung, Taiwan

Chia-Wei Tsai
Department of Computer
Science and
Information Engineering,
National Taichung University
of Science and Technology,
Taichung, Taiwan

Editorial Office
MDPI
St. Alban-Anlage 66
4052 Basel, Switzerland

This is a reprint of articles from the Special Issue published online in the open access journal *Electronics* (ISSN 2079-9292) (available at: https://www.mdpi.com/journal/electronics/special_issues/IoE_wireless).

For citation purposes, cite each article independently as indicated on the article page online and as indicated below:

Lastname, A.A.; Lastname, B.B. Article Title. *Journal Name* **Year**, *Volume Number*, Page Range.

ISBN 978-3-0365-8482-9 (Hbk)
ISBN 978-3-0365-8483-6 (PDF)
doi.org/10.3390/books978-3-0365-8483-6

Contents

About the Editors

Gwo-Jiun Horng

Gwo-Jiun Horng received his Ph.D. in Information Engineering from National Chenggong University, Taiwan, in 2013. He is currently a Professor in the Department of Computer Science and Information Engineering, Southern Taiwan University of Science and Technology. Prior to this, he served as an Assistant Professor in Cheng Shiu University, National Kaohsiung Marine University and Southern Taiwan University of Science and Technology. His research interests include AIoT, IoE, mobile service and computing, and wireless networks. In recent years, he has actively participated in academic activities, published over 100 papers in conferences and SCI, SCIE, and SSCI journals such as *Mobile Networks and Applications, Journal of Supercomputing, Computers and Electrical Engineering, ACM/Springer Mobile Networks and Applications*, and *IEEE Transactions on Intelligent Transportation Systems*. He also serves reviewers and editors in related journals. He has received awards including the Smart Transportation Paper Award and National Innovation Award and was sponsored by the Ministry of Science and Technology, Academia Sinica.

S.T. Aripriharta

Aripriharta received his Ph.D. from the National Kaohsiung University of Science and Technology, Kaohsiung, Taiwan, in 2017. Currently, he is an Associate Professor at the Faculty of Engineering, Department of Electrical Engineering, State University of Malang, Indonesia. His research interests focus on low-power electronics converters for biomedical IoT/wearable devices such as self-powered converters, energy harvesting, etc. He has published more than 100 publications, 10 patents, 20 copyrights and 5 books.

He also is a member of professional academic organizations such as IEEE, IAENG, IMRCS, and Fortei Reg 7.

Yao-Tung Tsou

Yao-Tung Tsou received a Ph.D. degree from the Department of Electrical Engineering, National Taiwan University, Taipei, Taiwan. He joined the Research Center for Information Technology Innovation (CITI), Academia Sinica, in 2013, as a Research Assistant. He is currently an Associate Professor at the Department of Communications Engineering, Feng Chia University, Taichung, Taiwan. He is also a Consultant at Lin Dan Technology Inc. and Swiss Innovation Valley. His research interests include Emergent Memory (STT-MRAM) Security System Application, Mobile Network Computing and Security, Cyber Security, and Embedded System Development and Application. He received the Magnetism Research Prize from the Taiwan Association for Magnetic Technology and the Best Paper Award in Taiwan Internet Seminar and AII2022 Taiwan Innovation and Invention Application Seminar. Dr. Tsou has published papers in conferences and SCIE journals such as *IEEE Transactions on Services Computing, Electronics, IEEE Access and Computer Communications*.

Chia-Wei Tsai

Dr. Chia-Wei Tsai is an Assistant Professor at the Department of Computer Science and Information Engineering at the National Taichung University of Science and Technology. In 2011, he received his Doctoral degree in Computer Science and Information Engineering from National Cheng Kung University, Tainan, Taiwan. His primary research interests are quantum cryptography, quantum information, quantum network, and anomaly detection. He has published 68 SCI journal papers in these fields. Furthermore, in terms of practical experiments, he also possesses 6 years of experience as an engineer in machine learning applications.

Editorial

Advances of Future IoE Wireless Network Technology

Gwo-Jiun Horng

Department of Computer Science and Information Engineering, Southern Taiwan University of Science and Technology, Yongkang District, Tainan 710301, Taiwan; grojium@stust.edu.tw

1. Introduction

The Internet of Everything (IoE) is a concept that refers to the interconnectivity of various devices, objects, and systems, which can communicate and exchange data to enable intelligent decision making. Wireless networks are at the forefront of IoE and they continue to advance, meeting the growing demands of the digital age. In this article, we summarize ten recent research articles that highlight the advances in IoE wireless network technology.

This collection of articles published in *Electronics* encompasses a wide range of topics related to emerging technologies. The articles include research on vendor-managed inventory mechanism based on the SCADA of the Internet of Things, in-memory computing architecture for a convolutional neural network, face prediction system for missing children in a smart city safety network, anomaly electricity usage behavior in residence using autoencoder, cloud-edge-smart IoT architecture for speeding up the deployment of neural network models, generative adversarial network and diverse feature extraction methods to enhance the classification accuracy of tool-wear status, classifying conditions of speckles and wrinkles on the human face using a deep learning approach, repetition with learning approaches in massive machine-type communications, GDPR personal privacy security mechanism for smart home systems, and the development of an autonomous vehicle training and verification system. These articles demonstrate the rapid advances being made in the field of electronics and highlight the potential for these technologies to impact on our daily lives.

In recent years, advances in technology have led to many exciting developments in the field of electronics. From smart city safety networks to autonomous vehicle training systems, researchers are constantly working on innovative solutions to complex problems. In this summary, we will discuss some of the latest research articles published in the *Electronics* journal.

2. Brief Description of the Published Articles

First, Kao and Chueh [1] proposed a vendor-managed inventory mechanism based on the SCADA of the Internet of Things Framework. This mechanism allows for vendors to monitor and manage the inventory levels of their customers in real time, enabling an efficient supply chain management.

Second, Huang et al. [2] presented an in-memory computing architecture for a convolutional neural network based on the spin orbit torque MRAM. This architecture improves the performance of the neural network while reducing energy consumption.

A study by Wang et al. [3] focus on the development of a face prediction system for missing children in a smart city safety network. The system uses deep learning techniques to predict the appearance of a missing child's face based on their current age and gender.

Another article by Tsai et al. [4] describe a method for detecting anomaly electricity usage behavior in residences using an autoencoder. By analyzing electricity usage patterns, the system can identify abnormal behavior that may indicate a potential safety or security issue.

Citation: Horng, G.-J. Advances of Future IoE Wireless Network Technology. *Electronics* **2023**, *12*, 2164. https://doi.org/10.3390/electronics12102164

Received: 6 May 2023
Accepted: 8 May 2023
Published: 9 May 2023

Hsu et al. [5] propose a cloud-edge-smart IoT architecture for speeding up the deployment of neural network models using transfer learning techniques. The proposed architecture allows for the efficient deployment of machine learning models on edge devices, reducing the need for expensive cloud infrastructure.

Chen et al. [6] present an application of generative adversarial networks (GANs) and diverse feature extraction methods to enhance the classification accuracy of tool-wear status. The study demonstrates the effectiveness of GANs in generating high-quality samples for training machine learning models.

An article by Chang and Tsai [7] focuses on the classification of conditions of speckles and wrinkles on the human face, using a deep learning approach. The proposed method achieves a high accuracy in identifying these conditions, which can be useful in cosmetic and medical applications.

Chen et al. [8] discuss the use of repetition with learning approaches in massive machine-type communications. The study proposes a framework for training machine learning models in resource-limited environments, such as those found in Internet of Things (IoT) devices.

Jhuang et al. [9] present a GDPR personal privacy security mechanism for smart home systems. The proposed mechanism provides enhanced security and privacy protections for personal data in smart home environments.

Finally, Wu et al. [10] describe the development of an autonomous vehicle training and verification system for teaching experiments. The system allows for the safe and efficient training of autonomous vehicles in real-world scenarios.

3. Conclusions

In conclusion, these articles demonstrate the wide range of applications for electronics in modern society. From machine learning and deep learning techniques, to smart city safety networks and autonomous vehicles, researchers are constantly pushing the boundaries of electronics technology possibilities.

Conflicts of Interest: The author declares no conflict of interest.

References

1. Kao, C.; Chueh, H. A Vendor-Managed Inventory Mechanism Based on SCADA of Internet of Things Framework. *Electronics* **2022**, *11*, 881. [CrossRef]
2. Huang, J.; Syu, J.; Tsou, Y.; Kuo, S.; Chang, C. In-Memory Computing Architecture for a Convolutional Neural Network Based on Spin Orbit Torque MRAM. *Electronics* **2022**, *11*, 1245. [CrossRef]
3. Wang, D.; Tsai, Z.; Chen, C.; Horng, G. Development of a Face Prediction System for Missing Children in a Smart City Safety Network. *Electronics* **2022**, *11*, 1440. [CrossRef]
4. Tsai, C.; Chiang, K.; Hsieh, H.; Yang, C.; Lin, J.; Chang, Y. Feature Extraction of Anomaly Electricity Usage Behavior in Residence Using Autoencoder. *Electronics* **2022**, *11*, 1450. [CrossRef]
5. Hsu, T.; Wang, Z.; See, A. A Cloud-Edge-Smart IoT Architecture for Speeding Up the Deployment of Neural Network Models with Transfer Learning Techniques. *Electronics* **2022**, *11*, 2255. [CrossRef]
6. Chen, B.; Chen, Y.; Loh, C.; Chou, Y.; Wang, F.; Su, C. Application of Generative Adversarial Network and Diverse Feature Extraction Methods to Enhance Classification Accuracy of Tool-Wear Status. *Electronics* **2022**, *11*, 2364. [CrossRef]
7. Chang, T.; Tsai, M. Classifying Conditions of Speckle and Wrinkle on the Human Face: A Deep Learning Approach. *Electronics* **2022**, *11*, 3623. [CrossRef]
8. Chen, L.; Ho, C.; Chen, C.; Liang, Y.; Kuo, S. Repetition with Learning Approaches in Massive Machine Type Communications. *Electronics* **2022**, *11*, 3649. [CrossRef]
9. Jhuang, Y.; Yan, Y.; Horng, G. GDPR Personal Privacy Security Mechanism for Smart Home System. *Electronics* **2023**, *12*, 831. [CrossRef]
10. Wu, C.; Wu, Y.; Liang, Y. The Development of an Autonomous Vehicle Training and Verification System for the Purpose of Teaching Experiments. *Electronics* **2023**, *12*, 1874. [CrossRef]

Article

A Vendor-Managed Inventory Mechanism Based on SCADA of Internet of Things Framework

Chang-Yi Kao [1] and Hao-En Chueh [2,*]

[1] Department of Computer Science & Information Management, Soochow University, Taipei City 100006, Taiwan; edenkao@scu.edu.tw
[2] Department of Information Management, Chung Yuan Christian University, Taoyuan City 320314, Taiwan
* Correspondence: hechueh@cycu.edu.tw

Citation: Kao, C.-Y.; Chueh, H.-E. A Vendor-Managed Inventory Mechanism Based on SCADA of Internet of Things Framework. *Electronics* **2022**, *11*, 881. https://doi.org/10.3390/electronics11060881

Academic Editor: Francisco Falcone

Received: 9 February 2022
Accepted: 7 March 2022
Published: 10 March 2022

Abstract: In recent years, with the rise of the Internet of Things (IoT) and artificial intelligence (AI), intelligent applications in various fields, such as intelligent manufacturing, have been prioritized. The most important issue in intelligent manufacturing is to maintain a high utilization rate of production. On the one hand, for maintaining high utilization, the production line must have enough materials at any time; on the other hand, too many materials in stock would greatly increase the operating cost of the factory. Therefore, maintaining sufficient inventory while avoiding excessive inventory is an important key issue in intelligent manufacturing. After the factory receives the order, it would issue the manufacturing order to the production line for manufacturing. The capacities of different production lines are different. If the Supervisory Control And Data Acquisition (SCADA) system based on the IoT framework can be used to monitor the capacity of each production line, in addition to estimating the capacity, the usage of key materials can also be accurately estimated through AI; when the quantity of key materials is below the safety stock, the manufacturer can actively notify the supplier and request for replenishment. This is a Customer-to-Business (C2B) safety stock management model (i.e., the vendor-managed inventory, VMI), which combines AI and IoT. In particular, in the case of consumer electronics, because their life cycles are short and they are vulnerable to market fluctuations, the manufacturer must adjust the production capacity. This study will propose to construct a SCADA system based on the IoT, including the capacity of the production line, materials inventory, and downstream order requirements, and use the Artificial Neural Network (ANN) to accurately predict inventory requirements. In this study, through the factory, a SCADA system based on AI and IoT will be constructed to monitor the factory's manufacturing capacity and predict the product sales of downstream manufacturers, for the purpose of facilitating the analysis and decision-making of safety stock. In addition to effectively reducing the inventory level, in essence, the purpose of this study is to enhance the competitiveness of the overall production and sales ecosystem, and to achieve the goal of digital transformation of manufacturing with AI and IoT.

Keywords: Internet of Things; SCADA; Customer-to-Business; vendor-managed inventory; Artificial Neural Network; digital transformation

1. Introduction

Intelligent manufacturing is an important application under Industry 4.0. In particular, after the outbreak of the COVID-19 pandemic, the monitoring system can be quickly set up through Artificial Intelligence Plus the Internet of Things (AIoT), and the management can be conducted remotely, for the purpose of reducing the cases of human-to-human transmission. Specifically, in response to the rise of the Internet of Everything (IoE), factories generally construct SCADA systems based on the IoT. This study proposes a safety stock management model (i.e., the VMI), with a view to improving the utilization and the on-time delivery of the factory itself. In addition, the AI prediction is conducted by interfacing with the product sales data of downstream manufacturers. Thus, the consumption of key materials and inventory requirements can be effectively estimated; when necessary, the manufacturer can actively report the quantity demanded to upstream key material suppliers, so that they can quickly provide sufficient inventory. This is also the intelligent application of intelligent factories in Industry 4.0: C2B safety stock services that integrate the value chain.

In the past, the production mode of the factory was that after the retail industry placed a purchase order to the manufacturing industry, the factory would schedule the production line for production, then issue the manufacturing order, and place the orders for materials separately, and the operation process was all manual operation. However, with the popularization of AI and IoT technologies, the C2B model has also emerged. C2B stresses the customer engagement business model, and the customer engagement C2B model would take the application of AIoT as the implementation model/approach. In the C2B model, firstly, the sales data of downstream manufacturers would be collected from the factory side through IoT services, which would be combined with SCADA data for analysis. Then, AI is used to estimate the usage of key materials and indicate whether they are sufficient or not (the notice is sent to upstream manufacturers or connected to upstream manufacturers' MES/SCM systems). This allows us to achieve a management model of key materials safety stock, improving the on-time delivery of production lines, and implementing the C2B service model that integrates downstream manufacturers. In this study, under the model of C2B integrated manufacturing order arrangement and management, SCADA is used to acquire information on the production line capacity, and combined with the sales prediction of downstream manufacturers, the usage of key materials and parts can be accurately estimated, thus practicing the management mode of C2B safety stock of key materials.

2. Literature Review

In the production model of consumer electronics, the manufacturing industry takes orders and then the factory manufactures the consumer electronics. However, in the future, the production characteristics of electronics will gradually move towards a trend of small-quantity but diverse electronics, which requires the ability to adjust the capacity of the production line with high flexibility. In the manufacturing industry, it is a standardized process to manufacture products after taking orders, for the purpose of delivering the products on time. In particular, the key raw materials for the products are generally stocked, because if there are no key materials in the production process, the production will be seriously delayed. However, the procurement specialists of different factories procure raw materials from the supply chain partners. In addition to long procurement times and slow response times at the supply side, there is no standard for raw material preparation quantities. These problems may cause excess inventory or insufficient material preparation, thus increasing the cost of the supply chain. The manufacturing industry has also begun to assist production management through various information and communication technologies (ICTs): predicting the quantity of materials demanded based on the orders of downstream manufacturers, which is an application that uses sales prediction to assist in estimating capacity requirements. In the past, the prediction model was mainly manual, which also resulted in the application effect of prediction in most enterprises being lower than expected [1]. Generally speaking, the manufacturing industry cannot grasp the sales status of downstream manufacturers. If the quantity of key materials cannot be estimated, it will cause the hoarding or shortage of key materials. The hoarding of materials would increase the cost of the enterprise, while the shortage of materials would make the enterprise unable to deliver the products on time. Thus, the enterprise would lose its competitiveness [2].

The production processes in the manufacturing industry consist of such steps as order taking, material requirement exploration, material and part management, material procurement, factory production, transportation management, and delivery. However, from the perspective of the strategy of the entire product value chain, the manufacturing industry responsible for production in the value chain should establish close integration with upstream suppliers and downstream customers to form a collaborative supply value chain. Collaborative Planning, Forecasting, and Replenishment (CPFR) [3,4] is a value chain collaboration model. CPFR focuses on efficiency, so the practice of Just-In-Time (JIT) is promoted and discussed [5–7]. JIT can reduce costs and waste, while it can increase flexibility and many interactions [8–10], such as delivery time and delivery quantity.

In addition, from the perspective of the manufacturing industry, the manufacturing industry is facing the trend of market globalization. The manufacturing industry does everything possible to reduce the costs of production and product development. Therefore, predicting the upstream material quantity based on the downstream orders is a collaboration that saves manufacturing costs, and it is also regarded as a C2B [11,12] model. This is particularly true for consumer electronics due to the high market volatility. Furthermore, the degree of digitalization of the manufacturing industry is not high, the basic environment is still in great demand, and the establishment rate of intelligent software is not high. Therefore, the top priority for using CPFR in the manufacturing industry is to transform the traditional supply chain into digital and intelligent information and communication services [13,14], interface the systems, and use the joint prediction model to predict the material inventory [15], with a view to improving the efficiency of collaboration.

In terms of safety stock management, some studies have used the PHLX Semiconductor Sector Index to predict the safety stock [16], which is a method of analysis and prediction based on the financial index. There are also related academic studies that use various indicators to analyze the material inventory, which is taken as a management model, but this mode is difficult to apply to the production of products that have nothing to do with financial indicators. In addition, the preparation time and delivery time of the order, the delivery time of the goods, and the preparation time of the goods are known as the lead time of the order, and the lead time can also be used as an estimation model for safety stock [17,18]. Analyzing the safety stock from the supplier's operation processes, using the relationship between industry indicators and demand [19], and so on, are all related research topics within safety stock management. The tools used include MACD [20], calculation of the demand index (DI) [19], and exponential moving average (EMA) [21].

This study proposes a mechanism based on CPFR to predict the safety stock in light of the orders or sales data at the downstream of the value chain, the utilization of the production line, and the usage of key materials. In this study, the basis for safety stock is sales prediction. With respect to a literature review on the studies on sales prediction, some scholars proposed the grey correlation analysis, in which the analysis is conducted by combining with the neural network [22]; in addition, [23,24] other scholars also proposed a new prediction model using the evolutionary neural network, the effect of which is better than that of the traditional neural network. Compared with the ARIMA prediction method [25], the fuzzy logic rules [26–28], as well as the inference mechanism [29–31] and the statistical model, the neural network does not need to specify a specific function type, and the data are not limited to a certain statistical distribution assumption, thus having more application space in the prediction. The data used for these analyses will be sales data at the downstream of the value chain, the data collected by IoT devices in factories, and the data from SCADA [32–34] monitoring systems. There will be no shortage of key materials under the safety stock prediction model of this study.

3. Proposed Architecture and Method

In this section, the overall framework of safety stock, the pre-processing of data analysis, and the prediction model will be explained.

3.1. System Framework

The SCADA system is important and essential in improving the production yield, digital transformation, and data retrospect of the manufacturing industry, and enhancing its competitive strengths, etc. However, in a traditional SCADA system, due to factors such as the fact that the consoles on the production line cannot provide the Application Programming Interface (API) and it is not easy to implement function expansion and subsequent maintenance, it is difficult for small-sized manufacturing industries to accept the traditional SCADA framework. Therefore, the construction of a cloud-based SCADA system, suitable for traditional small- and medium-sized manufacturing industries, would help the manufacturing industry to conduct operations management. To take the framework of the Internet of Things as the foundation, i.e., to transmit the data sensed and collected by the Internet of Things to the application layer through the network (i.e., the cloud-based SCADA system), is to improve the traceability of production data. The method proposed in this study can estimate the production capacity based on production data and, combined with the data of downstream manufacturers, conduct sales predictions. Predicting the required inventory based on the AI analysis of the data, the production line can maintain a high utilization rate.

The overall framework proposed in this study consists of two parts, as shown in Figure 1. The first part is the cloud-based solution for SCADA; the purpose of the SCADA solution in this study is to collect production process data in order to estimate production capacity, which includes the required IoT hardware and the corresponding real-time monitoring system. The second part is to intelligently analyze and estimate the required key materials by integrating the sales data of the suppliers, so as to provide the purchase reference for safety stock.

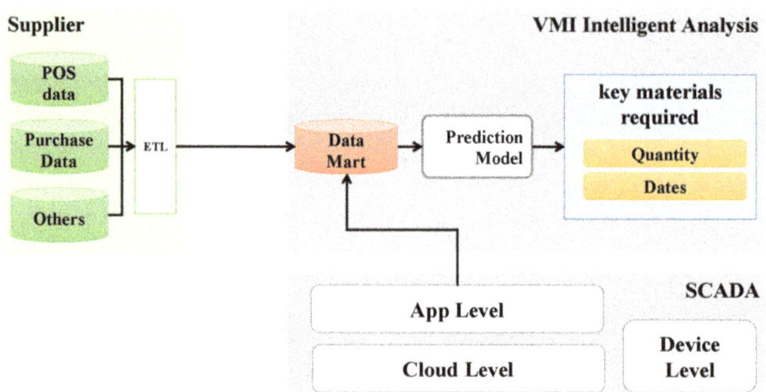

Figure 1. The architecture based on the SCADA of Internet of Things framework.

The first part is the cloud-based solution for SCADA. At the device level, the online IoT sensing device of production would transmit the data back to the SCADA system through the client data agent application, which would be set up on the gateway (GW). In order to identify which networked device senses and collects the data, the client data agent application has a simple pairing mechanism. In addition, in order to more easily manage the data collected by the client data agent application at the device level, the SCADA application includes authentication and behavior monitoring, etc. As shown in Figure 2, the factory side uses a variety of IoT devices for monitoring. Various sensing devices have different signal sources and data formats, different network addresses, and even different temperature units (Fahrenheit/Celsius). Previously, to solve this problem, it was necessary to write customized reading methods and control instructions for each IoT device on the GW side and convert the data format correctly. In this study, it is designed to standardize various signals and data formats and interface the signals through agents. The work items related to the device agent application interface are as follows:

- Analyze the device control signals and data signals (sequence signals). Develop the application for processing the underlying serial communication at the firmware layer, read different device communication interfaces according to different serial communication properties, and process control signals and data signal sequences, as well as maintain the status of intelligent devices, record the occurrence of events, and process error exceptions.

- Normalize the read and write formats and customized instructions, and design the data sheet. Design the standard data/instruction data format. Textualize the data read content. Do not judge the data content, but provide them to the upper level for processing. For the received instructions, decompose them in accordance with the specifications, and transmit them to the corresponding signal control layer.
- Determine the GW intermediary data format. Implement the upper layer data processing and control instruction logic, and convert the standard instruction into the corresponding address of each intelligent device. Then, provide the upper layer application, which can, through function calls, easily perform device setting, device testing, data reading, status notification, control instruction transmission, and instruction execution result return, without considering the actual corresponding address of the device.

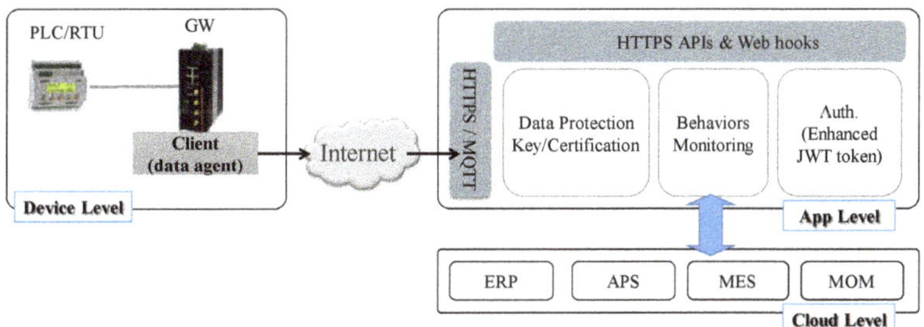

Figure 2. The SCADA system.

In the upsurge of the COVID-19 pandemic, there are still hackers who continue to invade information systems, which causes security concerns. If the centralized authorization mechanism of the information system is used for MQTT devices, the OAuth-based framework can be used. OAuth 2.0 supports decoupling authorization servers from resource servers (such as MQTT servers). When one uses OAuth 2.0, the client would provide its credentials to the authorization server, which then performs authentication checks and returns the access token that allows access to the resources, and then the access token is used to connect with the MQTT server. The MQTT server typically validates the access token by communicating with the authorization server, and then grants access to the resources. Each IoT gateway itself is associated with an independent unique ID, which will be used to generate a registration pairing exclusively for this IoT gateway, so that the user can directly register this IoT gateway on the cloud platform when deployed in the actual application field. In the database of the cloud platform, the datagram of each IoT gateway, including device registration model application and data format conversion, will be processed in the database. Finally, the data will be synchronized to the Data Mart of VMI Intelligent Analysis through HTTPS APIs and web hooks for analysis.

3.2. Prediction Method

Manufacturers' factories have their own management methods. Most of them rely on on-site personnel to adjust the management methods in light of their experience. There are no data for specific improvement. It is also easy to cause instability in production capacity due to communication gaps, and it is impossible to effectively promise the delivery status to downstream customers. In this study, the data sensed by the IoT device, such as the material barcode, operator identification card, production machine number, and manufacturing order number, are transmitted back to the cloud-based SCADA system through the agent.

Then, the learning-to-rank method is used to predict the change in production capacity to estimate the quantity of key materials demanded for production. The eigenvalues of analysis are transmitted from the data collected by SCADA to the Data Mart. (Table 1)

Table 1. Characteristic data for estimating production capacity.

Data Type	Description	Effect [1]
Historical value of production capacity last week ... Historical value of production capacity 52 weeks ago	Use historical data as the benchmark for estimating production capacity	Depends on
Product remake ratio	High product remake ratio, and reduced production capacity load	Negative
Production lead time	Long production lead time, which affects production capacity	Negative
Machine utilization	High machine utilization will increase production capacity	Positive
Operator's productivity	The operator operates the machine, which will greatly affect the production capacity	Depends on

[1] Positive/negative effect on production capacity.

Because the factory SCADA has data records (log) for the materials and parts, the production capacity data of SCADA in this study are collected for 52 weeks, i.e., data covering a whole year. After estimating the production capacity of the week, an alert can be issued for the safety stock of key materials. The production capacity estimation method proposed in this study is based on the concept of AdaBoost, and a series of rough rules are weighted and combined to obtain highly accurate rules. Its advantages include the following: (1) it is easy to implement; (2) the classification accuracy is high. Its disadvantages include the following: (1) it is easily disturbed by noise, which is also a disadvantage of most sorting or classification algorithms; (2) the execution effect depends on the selection of weak classifiers. AdaBoost is most suitable for processing large-dimensional characteristic data, because it can generate detection results with high efficiency and high detectivity.

This study employs AdaBoost, which can dynamically adjust the ranking of data weights, turn the ranking problem into binary, and improve the part of updated data weights. The essence of weight is that the more the data are not classified correctly, the higher the weight of the data is. In other words, not only the order of data in the correct ranking but also the distance of data in the correct ranking are considered. Then, the data point is ranked first if the concordant is positive (representing the high ranking of the data point) and the data point has the smallest weight coefficient (the farthest from the center of gravity). The data point is ranked second if the concordant is positive and the weight coefficient is next (the next farthest away from the data point), and so on.

$$D_{t+1}\left(x_i, x_j\right) = \frac{D_t\left(x_i, x_j\right) \exp\left(1 - \alpha_t\left(1 - dd_t\left(x_i, x_j\right)\right)\right)}{Z_t}, if\ldots concordant\ldots, \qquad (1)$$

$$D_{t+1}\left(x_i, x_j\right) = \frac{D_t\left(x_i, x_j\right) \exp\left(\alpha_t\left(dd_t\left(x_i, x_j\right)\right)\right)}{Z_t}, if\ldots discordant\ldots \qquad (2)$$

When it is concordant: $if, d_t\left(x_i, x_j\right) * d^*\left(x_i, x_j\right) > 0$
When it is discordant: $if, d_t\left(x_i, x_j\right) * d^*\left(x_i, x_j\right) < 0$

$$where, dd_t\left(x_i, x_j\right) = \left|d_t\left(x_i, x_j\right) - d^*\left(x_i, x_j\right)\right|$$
$$d_t\left(x_i, x_j\right) = h_t(x_i) - h_t(x_j)$$
$$d^*\left(x_i, x_j\right) = h^*(x_i) - h^*(x_j)$$

At the same time, the pair difference of the ideal ranking function *h** is considered, and then the difference from the subtraction represents the similarity between the weak learner and the correct ranking. The two cases of concordant and discordant are discussed separately. The learning-to-rank method is used to find out which week's production capacity in the 53 weeks is the closest to that of the estimated week, and the closest one is ranked first. This study uses the ranking similarity as a means to estimate the production capacity. However, there is still a lack of important data to estimate the safety stock of key materials, i.e., the order quantities of downstream manufacturers. However, usually, the order quantities cannot be estimated, so this study estimates how much production volume may be required in the future from the sales prediction of downstream manufacturers, and then calculates the key materials that may be required. Because the SCADA production capacity estimation collects 52-week historical data, the sales data are also analyzed using 52-week data. (Table 2)

Table 2. Characteristic data for sales prediction.

Data Type	Description	Data Source
Historical value of sales volume last week		
...	Used as the benchmark for estimating production capacity	POS data of downstream manufacturers
Historical value of sales volume 52 weeks ago		
Promotion information	Are there any special offers this week?	
Traditional festivals covered	Whether they include New Year's Day, Mother's Day, and Christmas	Manual seeding
Winter and summer vacations	Whether it is winter or summer vacation	

Data analysis requires data normalization to compress the data. In other words, dimensionality reduction is performed through PCA, and the eigenvalues with a high correlation with the prediction target are reserved as the dependent variable of sales prediction. Among the many eigenvalues, eliminating the variables with low correlation to sales is called data preprocessing. After data preprocessing, the sales data, discount data, traditional festivals covered, and winter and summer vacations in the previous week, previous 4 weeks, previous 8 weeks, and previous 12 weeks are important variables, so they are selected as the eigenvalue variables of the prediction model.

In this study, the prediction model employs the Back-Propagation Neural Network (BPN) of artificial intelligence. BPN is used by many scholars for prediction analysis. BPN includes an input layer, a hidden layer, and an output layer. The model uses the hidden layer to convert the data of the input layer into the nonlinear function, and then the hidden layer is converted again in the output layer. BPN has been used in research in other fields. The steps used in this study are briefly described as follows:

Training stage hidden layer:

$$net_h = \sum_i W\_xh_{ih} \cdot X_i - \theta\_h_h \tag{3}$$

$$H_h = f(net_j) = \frac{1}{1 + \exp^{-net_h}} \tag{4}$$

and $\Delta W\_xh_{ih} = \eta \delta_h X_i + \alpha \cdot \Delta W\_xh_{ih}$, $\Delta \theta\_h_h = -\eta \delta_h + \alpha \cdot \Delta \theta\_h_h$.

The output is $W\_hy_{hj} = W\_hy_{hj} + \Delta W\_hy_{hj}$, $\theta\_y_j = \theta\_y_j + \Delta \theta\_y_j$.

H_h is the output vector, ΔW is *Wij*, which mimics the strength of the connection between the *i*-th and *j*-th neurons, and δ is the amount of difference between the processing unit connected to W and the upper-level processing unit.

In the validation stage, the error function of the network is minimized, and the learning quality of the model is generally adjusted by using the formula error function. In this study, the execution result of the proposed prediction model is compared with the usage of key materials in the production line capacity prediction. After deducting the usage of key materials of the order in the prediction model of the deep learning model from the usage of key materials, if it is lower than the threshold value, it indicates that the key materials will soon be short, i.e., the stock is unsafe, and the key materials shall be ordered from the suppliers of the key materials; on the contrary, the key materials are still within the scope of safety stock.

4. Experience Result

After the rise of the Internet of Things, many emerging electronics are moving towards customized production. Factory production is characterized by a small quantity/variety and high added value, which requires high flexibility to adjust the production line capacity. After the order is established, the electronics will be manufactured according to the manufacturing order, and the factory will receive the order, customize the electronics, issue the manufacturing order, and deliver the electronics. In the future, consumer electronics will require a highly customized production strategy; the management of key materials and parts required for production is very important, especially the safety stock management of key materials.

This study takes consumer electronics as the experimental subject, uses the IoT to transmit production data to the SCADA cloud-based application system under the aforementioned framework, and then extracts and analyzes the data from the SCADA cloud-based application system. The weekly production capacity data and product remake ratio from 1 week to 52 weeks are obtained from the Work In Process (WIP) tracking system of its single production line, the production lead time is obtained from the automatic Material Control System (MCS), and the machine utilization and operator's productivity are obtained from the Engineering Data Collection (EDC) system. After the expected production capacity is finally estimated, the quantities of key materials in the Material Requirements Planning (MRP) system are compared. Each group of data in this experiment is on a weekly basis, with a total of 106 groups of data (which means the data in 106 weeks). There are a total of 53 groups of data (the data in 53 weeks) in one year, so there are 2 years of data. Among them, the data of a complete year are used as the training data, while the other 53 groups of data are used as the validating data. In order to estimate the production capacity through the ranking, we need to have a group of estimated data, and then compare this group of data with the 53 groups of data for the ranking. This group of estimated data will be set by the experts in light of their experience. In the future, the estimated value can be given by the factory director or manager under actual circumstances. The histogram in Figure 3 is the training data, the triangle legend of the line chart is the estimated production capacity using the ranking method of Learning-To-Rank, and the circle legend of the line chart is the validating data. The method proposed in this study can be observed from Figure 3: (1) The estimated production capacity is roughly in line with the trend of the validating data; (2) The production capacity peaks in the first and fourth quarters, and troughs in the third quarter. In the section of data validation, the Mean Absolute Percentage Error (MAPE) and Mean Absolute Error (MSE) are used for comparing the prediction error value. In other words, the model trained with the data of the previous year is used to compare the prediction error value with the data of the second year. Judging from the errors in the table below (Table 3), for estimating production capacity, the prediction model of Learning-To-Rank is feasible.

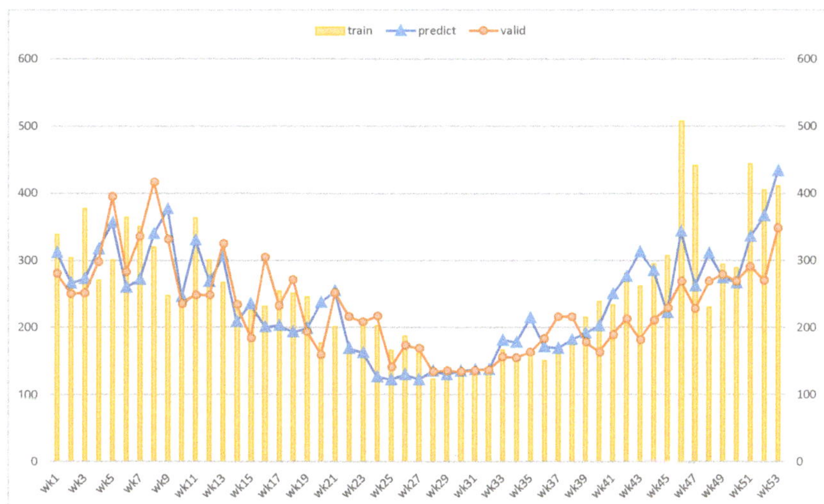

Figure 3. Validation and predicted data.

Table 3. Comparison of the models for estimating production capacity.

Prediction Model	MSE		MAPE	
	Training Data	**Validating Data**	**Training Data**	**Validating Data**
Time Series Regression	68.36%	60.51%	5.26%	5.94%
This Study	58.66	50.26	4.05%	5.03%

After extracting the data from the SCADA cloud-based application system and estimating the production capacity, the sales of downstream manufacturers are predicted. Most of the downstream manufacturers are the brand players or retailers. The brand players provide products to the retail terminals, while the retailers will sell directly. In the sales prediction, the overall sales will be predicted. In other words, the brand players and retailers are integrated:

Retailers = $(R_1, R_2, R_3, \ldots, R_n)$.

R_1 All eigenvalue data $F_{R1} = (F_{R1,1}, F_{R1,2}, F_{R1,3}, \ldots, F_{R1,n})$.

R_2 All eigenvalue data $F_{R2} = (F_{R2,1}, F_{R2,2}, F_{R2,3}, \ldots, F_{R2,n})$. and so on.

All downstream manufacturers training data set $D_f = (F_1, F_2, F_3, \ldots, F_n)$

$$D_f = \begin{bmatrix} F_{R1,1} + F_{R2,1} + F_{R3,1} \ldots F_{Rn,1} \\ F_{R1,2} + F_{R2,2} + F_{R3,2} \ldots F_{Rn,2} \\ \ldots \\ F_{R1,m} + F_{R2,m} + F_{R3,m} \ldots F_{Rn,m} \end{bmatrix} \tag{5}$$

After collecting the characteristic data of sales, this study uses the aforementioned neural network for training, and compares the predicted value and the order value with the sales quantity, to observe the effect of the prediction model. The order value here refers to the total quantity ordered from the manufacturers.

1. The ratio of purchase orders to sales
2. The ratio of forecasting to sales

It can be found from Figure 4 that, with respect to the sales prediction for one year (a total of 53 weeks) that is conducted according to this model, the neural network predication of this method is completely consistent with the sales trend. It is validated that the proposed prediction model can provide effective sales prediction.

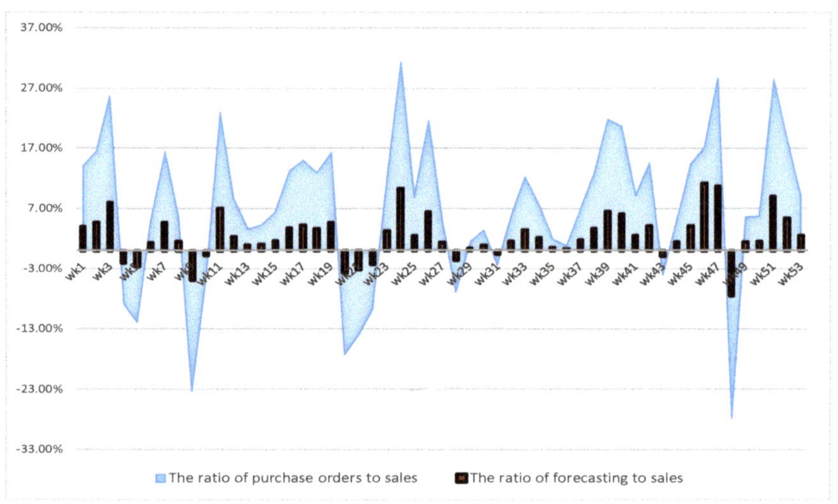

Figure 4. Comparison of prediction methods.

The histogram is the percentage of the difference between the prediction and sales in the proposed model, while the zone chart is the percentage of the difference between the order and sales. The closer the percentage of the difference is to 0, the more it can reduce the stock of the brand player or retailer, and the more accurate the manufacturing demand that can be provided to the factory will be. As can be seen from the data in the diagram, for the proposed model, the difference between the predicted value and the sales value is smaller.

The research method has been able to effectively estimate the production capacity and predict the sales of the downstream brands and the retail industry. It is reasonable to assume that after the sales performance improves, there will be more orders to the factory. After the number of orders to the factory increases, the production capacity must be increased to meet the quantities of products in the corresponding orders. Then, the added production capacity will lead to a shortage of materials, especially key materials. We only discuss one key material here.

Taking the mobile phone as an example, the panel is the most important key material. Therefore, after deducting the quantity that can be produced by the production capacity from the quantity of products in the sales prediction, the result will be the output to be added. In other words, the key material may be in short supply. Then, the factory should use the API of the SCADA system to directly and automatically order the key material from the upstream material supplier, and request the upstream supplier to provide the key material, to ensure that the production capacity is uninterrupted and the high utilization rate of the production line is maintained. The histogram in Figure 5a is the estimated output based on the production capacity and the predicted sales, and the line chart is the result of the estimated production capacity minus the predicted sales, so it is the estimated remaining stock of key materials. It can be seen from Figure 5a that although the estimated remaining quantity of key materials per week remains within plus or minus 20%, it can be found from the accumulative histogram of remaining key materials in Figure 5b that there is a serious shortage of key materials in the second quarter. At this time, the shortage of key materials is the safety stock that the factory should replenish. Lewis (1982) stated that the model predictive ability of MAPE is good between 10% and 20%. Thus, the results of this study are good. The SCADA framework and analysis model in this study can identify the shortage in advance, so that the factory can replenish the safety stock as soon as possible, thus effectively improving the utilization of the production line.

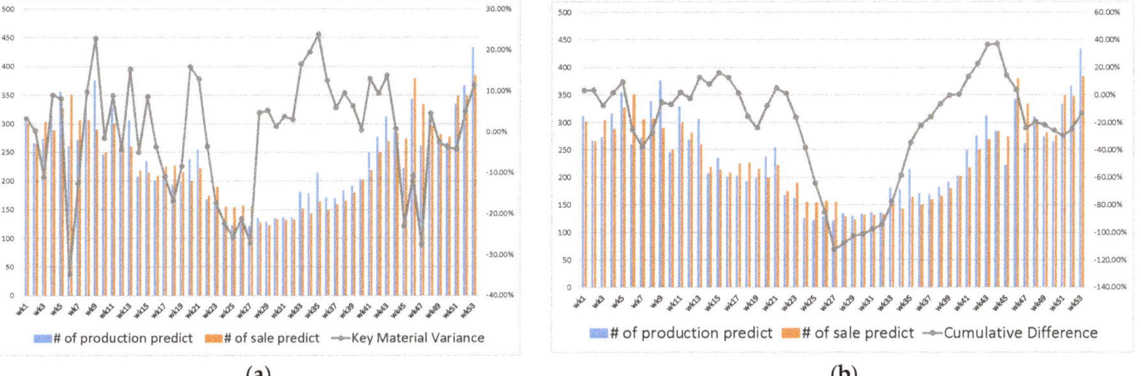

(a) **(b)**

Figure 5. Comparison of key materials safety stock under production capacity and sales prediction: (**a**) Difference between estimated production capacity estimates and sales prediction; (**b**) Accumulation of the difference between production capacity estimates and sales prediction.

The ARIMA model has been one of the most well-accepted methods in the study of time series forecasting. In recent years, BPN has also proven to be valuable for time series prediction applications, especially for non-linear models with very significant processing capabilities. This is because BPN processes the time series problem by predicting the $p + 1$ values from the data of the p value. The BPN only uses information on autoregressive and ignores information on moving average. Therefore, the model combination of LSTM may be a better approach for the prediction study related to time series problems.

5. Conclusions

The Internet of Things (IoT) has now evolved into the Internet of Everything (IoE), which makes all applications more diverse and efficient. The typical application of the IoT is the intelligent factory: the sensing object of the factory will store the sensed data through the network to the application layer (i.e., the cloud-based SCADA system) for monitoring. However, in addition to monitoring, the application of the IoT in the factory can also result in many value-added applications, such as the intelligent analysis of safety stock. In this study, the data, including the product remake ratio, production lead time, and operator's productivity, are stored in the cloud-based SCADA system through the IoT. The production capacity is estimated through the machine learning of a weak classifier. After collecting the sales data of downstream brand players and retailers, the neural network is used for training and sales prediction, and then the sales prediction and production capacity are compared. If the predicted sales volume is much larger than the current production capacity of the factory, it indicates that downstream manufacturers will increase the number of orders, and the factory is bound to improve the production capacity and add key materials, so as to manage the safety stock in this model. This study proposes a safety stock management model using the IoT and AI, which integrates downstream demand and its own production capacity for analysis through the Internet service, and provides a reference for the safety stock of key materials. Meanwhile, this is a C2B model of "customers participating in enterprise operation and management", and it is also a model for enterprises to use the Internet of Things to develop innovative services and achieve the goal of the digital transformation of manufacturing with AI and IoT. Thus, the issue of VMI should be temporal. The future directions of related research would be to compare time series forecasting method models, recurrent neural networks, or time series regarding their performance with the proposed methods.

Author Contributions: Conceptualization, C.-Y.K.; methodology, C.-Y.K.; validation, C.-Y.K. and H.-E.C.; formal analysis, C.-Y.K. and H.-E.C.; investigation, C.-Y.K. and H.-E.C.; data curation, C.-Y.K. and H.-E.C.; writing—original draft preparation, C.-Y.K.; writing—review and editing, C.-Y.K. and H.-E.C.; visualization, C.-Y.K.; supervision, C.-Y.K. and H.-E.C. All authors have read and agreed to the published version of the manuscript.

Funding: This research received no external funding.

Conflicts of Interest: The authors declare no conflict of interest.

References

1. Adebanjo, D.; Mann, R. Identifying Problems in Forecasting Consumer Demand in the Fast Moving Consumer Goods Sector. *Benchmark. Int. J.* **2000**, *7*, 223–230. [CrossRef]
2. Venkatraman, N.; Ramanujam, V. Measurement of business performance in strategy research: A comparison of approaches. *Acad. Manag. Rev.* **1986**, *11*, 801–815. [CrossRef]
3. Panahifar, F.; Byrne, P.J.; Heavey, C. A hybrid approach to the study of CPFR implementation enablers. *Prod. Plan. Control* **2015**, *26*, 1090–1109. [CrossRef]
4. Lin, R.H.; Ho, P.Y. The study of CPFR implementation model in medical SCM of Taiwan. *Prod. Plan. Control* **2014**, *25*, 260–271. [CrossRef]
5. Folinas, D.K.; Fotiadis, T.A.; Coudounaris, D.N. Just-in-time theory: The panacea to the business success? *Int. J. Value Chain. Manag.* **2017**, *8*, 171–190. [CrossRef]
6. Wang, D.; Chen, Y.J.; Chen, D. Efficiency optimization and simulation to manufacturing and service systems based on manufacturing technology Just-In-Time. *Pers. Ubiquitous Comput.* **2018**, *22*, 1061–1073. [CrossRef]
7. Zhou, B.H.; Peng, T. Scheduling the in-house logistics distribution for automotive assembly lines with just-in-time principles. *Assem. Autom.* **2017**, *37*, 51–63. [CrossRef]
8. Willis, T.; Huston, C. Vendor requirements and evaluation in a JIT environment. *Int. J. Oper. Prod. Manag.* **1990**, *10*, 41–50. [CrossRef]
9. Lawrence, J.J.; Lewis, H.S. Understanding the use of just-in-time purchasing in a developing country: The case of Mexico. *Int. J. Oper. Prod. Manag.* **1996**, *16*, 68–80. [CrossRef]
10. Schonberger, R.J.; Gilbert, J.P. Just-in-time purchasing: A challenge for U.S. industry. *Calif. Manag. Rev.* **1983**, *26*, 54–68. [CrossRef]
11. Liang, C.C.; Liang, W.Y.; Tseng, T.L. Evaluation of intelligent agents in consumer-to-business e-Commerce. *Comput. Stand. Interfaces* **2019**, *65*, 122–131. [CrossRef]
12. Zhang, X.Y.; Ming, X.G.; Liu, Z.; Yin, D. State-of-the-art review of customer to business (C2B) model. *Comput. Stand. Interfaces* **2019**, *132*, 207–222. [CrossRef]
13. Sagar, N. CPFR at Whirlpool Corporation: Two heads and an exception engine. *J. Bus. Forecast. Methods Syst.* **2003**, *22*, 3–8.
14. Simchi-Levi, D.; Kaminsky, P. *Designing and Managing the Supply Chain: Concepts, Strategies and Case Studies*, 3rd ed.; McGraw Hill Professional: New York, NY, USA, 2009.
15. Danese, P. Collaboration forms, information and communication technologies, and coordination mechanisms in CPFR. *Int. J. Prod. Res.* **2006**, *44*, 3207–3226. [CrossRef]
16. David, F.R. *Competing through Supply Chain Management: Creating Market-Winning Strategies through Supply Chain Partnership*; Springer: Berlin/Heidelberg, Germany, 1997.
17. Pyke, D.F.; Peterson, R.; Sliver, E.A. Inventory Management and Production Planning and Scheduling. *J. Oper. Res. Soc.* **2001**, *52*, 845.
18. Tony, W. *Best Practice in Inventory Management*, 3rd ed.; Elsevier Science Ltd.: Amsterdam, The Netherlands, 2002.
19. Gerald, A.W. *Frederick Hitschler, Stock Market Trading Systems: A Guide to Investment Strategy*, 1st ed.; Traders Press. Inc.: Cedar Falls, IA, USA, 1979.
20. Waheed, A. Analysis of Moving Average Convergence Divergence (MACD) as a Tool of Equity Trading at the Karachi Stock Exchange. *J. Financ.* **2013**, *20*, 78–92.
21. Wilson, A.R. Event triggered analog data acquisition using the exponential moving average. *IEEE Sens. J.* **2014**, *14*, 2048–2055. [CrossRef]
22. Chen, F.L.; Ou, T.Y. Grey relation analysis and multilayer function link network sales forecasting model for perishable food in convenience store. *Expert Syst. Appl.* **2009**, *36*, 7054–7063. [CrossRef]
23. Sun, Z.L.; Choi, T.M.; Au, K.F.; Yu, Y. Sales forecasting using extreme learning machine with applications in fashion retailing. *Decis. Support Syst.* **2008**, *46*, 411–419. [CrossRef]
24. Chen, Y.; Yao, Y. A multiview approach for intelligent data analysis based on data operators. *Inf. Sci. Int. J.* **2008**, *178*, 1–20. [CrossRef]
25. Lee, Y.S.; Tong, L.I. Forecasting time series using a methodology based on autoregressive integrated moving average and genetic programming. *Knowl. Based Syst.* **2011**, *24*, 66–72. [CrossRef]
26. Hilletofth, P.; Sequeira, M.; Adlemo, A. Three novel fuzzy logic concepts applied to reshoring decision-making. *Expert Syst. Appl.* **2019**, *126*, 133–143. [CrossRef]

27. Rajeswari, A.M.; Deisy, C. Fuzzy logic based associative classifier for slow learners prediction. *J. Intell. Fuzzy Syst.* **2019**, *36*, 2691–2704. [CrossRef]
28. Yadav, H.B.; Kumar, S.; Kumar, Y.; Yadav, D.K. A fuzzy logic based approach for decision making. *J. Intell. Fuzzy Syst.* **2018**, *35*, 1531–1539. [CrossRef]
29. Golemanova, E.; Golemanov, T.; Kratchanov, K. Comparative Study of the Inference Mechanisms in PROLOG and SPIDER. *TEM J. Technol. Educ. Manag. Inform.* **2018**, *7*, 892–901.
30. Araujo, D.A.; Hentges, A.R.; Rigo, S.J.; RighiSilva, R.R. Applying Parallelization Strategies for Inference Mechanisms Performance Improvement. *IEEE Lat. Am. Trans.* **2018**, *16*, 2881–2887. [CrossRef]
31. Buelens, B.; Burger, J.; van den Brakel, J.A. Comparing Inference Methods for Non-probability Samples. *Int. Stat. Rev.* **2018**, *86*, 322–343. [CrossRef]
32. Tao, L.; Siqi, Q.; Zhang, Y.; Shi, H. Abnormal Detection of Wind Turbine Based on SCADA Data Mining. *Math. Probl. Eng.* **2020**, *9*, 751. [CrossRef]
33. Gonzalez, E.; Tautz-Weinert, J.; Melero, J.J.; Watson, S.J. Statistical evaluation of SCADA data for wind turbine condition monitoring and farm assessment. *J. Phys. Conf. Ser.* **2018**, *1037*, 032038. [CrossRef]
34. Pandit, R.K.; Infield, D. SCADA-based nonparametric models for condition monitoring of a wind turbine. *J. Eng.* **2019**, *2019*, 4723–4727. [CrossRef]

 *electronics*

Article

In-Memory Computing Architecture for a Convolutional Neural Network Based on Spin Orbit Torque MRAM

Jun-Ying Huang [1], Jing-Lin Syu [2], Yao-Tung Tsou [2,*], Sy-Yen Kuo [1] and Ching-Ray Chang [3]

[1] Department of Electrical Engineering, National Taiwan University, Taipei 106, Taiwan; junying1995@gmail.com (J.-Y.H.); sykuo@ntu.edu.tw (S.-Y.K.)
[2] Department of Communications Engineering, Feng Chia University, Taichung 407, Taiwan; qaz7531213@gmail.com
[3] Quantum Information Center, Chung Yuan Christian University, Taoyuan 320, Taiwan; crchang@phys.ntu.edu.tw
* Correspondence: yttsou@fcu.edu.tw

Abstract: Recently, numerous studies have investigated computing in-memory (CIM) architectures for neural networks to overcome memory bottlenecks. Because of its low delay, high energy efficiency, and low volatility, spin-orbit torque magnetic random access memory (SOT-MRAM) has received substantial attention. However, previous studies used calculation circuits to support complex calculations, leading to substantial energy consumption. Therefore, our research proposes a new CIM architecture with small peripheral circuits; this architecture achieved higher performance relative to other CIM architectures when processing convolution neural networks (CNNs). We included a distributed arithmetic (DA) algorithm to improve the efficiency of the CIM calculation method by reducing the excessive read/write times and execution steps of CIM-based CNN calculation circuits. Furthermore, our method also uses SOT-MRAM to increase the calculation speed and reduce power consumption. Compared with CIM-based CNN arithmetic circuits in previous studies, our method can achieve shorter clock periods and reduce read times by up to 43.3% without the need for additional circuits.

Keywords: convolution neural network; computing in memory; processing in memory; distributed arithmetic; MRAM; SOT-MRAM

Citation: Huang, J.-Y.; Syu, J.-L.; Tsou, Y.-T.; Kuo, S.-Y.; Chang, C.-R. In-Memory Computing Architecture for a Convolutional Neural Network Based on Spin Orbit Torque MRAM. *Electronics* **2022**, *11*, 1245. https://doi.org/10.3390/electronics11081245

Academic Editor: Marco Vacca

Received: 7 March 2022
Accepted: 11 April 2022
Published: 14 April 2022

Publisher's Note: MDPI stays neutral with regard to jurisdictional claims in published maps and institutional affiliations.

1. Introduction

Machine learning (ML) models, such as convolutional neural networks (CNNs) and deep neural networks (DNNs), are widely used in real-world applications. However, neural network structures have also increased in size, causing a bottleneck in the Von Neumann accelerator architecture. More specifically, the CPU must retrieve data from memory before processing it and then transfer it back to memory at the end of the computation in a Von Neumann architecture. This leads to additional energy consumption during data transfer, which reduces the energy efficiency of computing devices [1]. Furthermore, limited memory bandwidth, high memory access latency, and long memory access paths limit inference speeds and cause substantial power consumption regardless of the performance of the logic circuit. However, in-memory computing can effectively overcome the bottlenecks in the Von Neumann architecture. The CIM architecture can achieve low memory access latency, parallel operation, and ultra-low power consumption, and close access to the arithmetic logic unit of the CIM architecture can overcome the bottlenecks of the Von Neumann architecture [2].

The error rate of visual recognition by CNNs declined from 28% in 2010 to 3% in 2016 to become better than the 5% error rate in manual (i.e., human) visual recognition [3]. CNNs have been integrated into embedded systems to solve image classification and pattern recognition problems. However, large CNNs may have millions of parameters and require up to tens of billions of operations to process an image frame [4]. Therefore, accelerating the convolution operation yields the greatest improvement in performance. Iterative processing of the CNN layers is a common design feature of CNN accelerators. However, the intermediate data are too large to fit on the chip's cache, and accelerator designs must, thus, use off-chip memory to store intermediate data between layers after processing. Due to the computational requirements of Internet of Things (IoT) and artificial intelligence (AI) applications, the cost of moving data between the central processing unit (CPU) and memory is a key limiter of performance.

CPU and GPU performance growth is approximately 60% per year; however, the performance increases in memory reach only up to 7% each year [5]. The data transfer rate in memory is insufficiently fast for the computational speed of the CPU; thus, the CPU is typically "data hungry." Although deep learning processor performance has grown exponentially, most power consumption occurs during the reading and writing of data. Thus, the efficiency of the accelerator has little effect on performance.

In the field of hardware design, the computing units, such as GPUs, CPUs, and isolated memory modules, are interconnected with buses; this design has entailed multiple challenges, such as long memory access latency, limited memory bandwidth, substantial energy requirements for data communication, congestion during input and output (I/O), substantial leakage power consumption when storing network parameters in volatile memory. Additionally, because the memory used for AI accelerators is volatile memory, data are lost if power is lost. Therefore, overcoming these challenges is imperative for AI and CNN applications.

To design a hardware CNN accelerator with improved performance and reduced energy consumption, CIM CNN accelerators [5–7] constitute a viable method to overcome the "CNN power and memory wall" problem; these accelerators have been researched extensively. The key concept of CIM is the embedding of logic units within memory to process data by leveraging inherent parallel computing mechanisms and exploiting higher internal memory bandwidth. CIM could lead to remarkable reductions in off-chip data communication latency and energy consumption. In the field of algorithm design, several methods have been proposed to break the memory wall and power wall; these include compressing pretrained networks, quantizing parameters, using binarization, and pruning. Additionally, Intel's Movidius Neural Compute Stick is a hardware NN accelerator for increased computing performance. In contrast, our approach is based on the MRAM CIM architecture, but the Movidius Neural Compute Stick is an onboard computing architecture. Compared to onboard computing architectures, MRAM CIM-based architectures significantly reduce the costs associated with data exchange between storage and memory. Our architecture has several key advantages, including non-volatility (no data loss in the absence of power), lower power consumption, and higher density. With the increasing demand for on-chip memory on AI chips, MRAM is emerging as an attractive alternative.

This paper follows the same assumptions in the existing works [5,8] and primarily focuses on methods of reducing hardware power consumption in edge computing without software algorithms. We designed a CIM CNN accelerator that is compatible with all the aforementioned algorithms without modifying the hardware architecture. Notably, we do not tackle the influence of slower peripherals to CNN.

Our contributions can be summarized as the follows:

- Integrate a DA architecture with the CIM to achieve faster speeds, fewer reads and writes, and lower power consumption.
- Optimize CNN operations and complete calculations in fewer steps.
- Integrate the DA architecture with the CIM and magnetic random access memory (MRAM) techniques to replace the original circuit architecture without off-chip memory. All calculations are performed on the cell array; thus, low latency can be achieved.
- Parallelize the CIM process using calculations in a sense amplifier to reduce power consumption and accelerate calculations.

The rest of this paper is organized as follows. Section 2 describes background and related work. We then describe the details of proposed architecture in Section 3. We provide the experimental process and results in Section 4. Finally, the conclusion is presented in Section 5.

2. Background and Related Work

2.1. CNN

CNN [9] is a combination of a feature extractor and a category classifier. The architecture uses shared kernel weights, local receptive fields, and spatial and temporal pooling to ensure invariance with respect to shift, scale, and distortion. Moreover, novel layers have been developed, such as the normalization layer and the dropout layer. CNN models typically have a feed-forward design; each layer uses the output of the previous layer as its input, and its calculation are output to the next layer. CNNs typically comprise three primary types of layers: the convolution (CONV) layer, the pooling layer, and the fully connected (FC) layer.

The convolutional layer is the main layer of a CNN. Each output pixel is connected to a local region of the input layer; this connection is called the receptive field. The receptive field can be defined as the window size of the region in the local input that produces a feature. These connections scan the entire image in the input feature map by extending a fixed-size window along the length and width of the entire image. The displacement of the window (i.e., the overlap of the receptive fields in both the height and width) typically has a value of 1 and is shared with the kernel weight. The processes of convolution is a 2D operation in which the shared kernel weight is multiplied element-wise with the corresponding receptive field. These element-wise operations require numerous executions of the multiplication and addition operations.

An input layer typically contains multiple channels, and the sum of all the channels is the result of the convolution. Pixel y at position (x, y) in the convolution result for n is given as follows.

$$sum = \sum_{i=0}^{N_i-1} \sum_{j=0}^{N_y-1} \sum_{k=0}^{K_x-1} w^{(n)}[i][j][k] \times in[i][x+j][y+k] \tag{1}$$

$$y^{(n)}[x][y] = f(sum + b[n]) \tag{2}$$

Therefore, if input data are extended to three dimensions of length, width, and depth, each 2D kernel must correspond to a depth.

The pooling layer can reduce the results of feature extraction but retain important features, typically reducing the image size by half. The pooling layer is generally placed following the convolutional layers. Average pooling and max pooling are two common pooling methods. In average pooling, the average value of the local field in each input feature map is calculated, whereas in max pooling, the maximum of the local field parameter is selected and pixels are output. Moreover, the number of output feature maps in the pooling layer must be equal to the number of input feature maps. Reducing the parameters can increase the efficiency of system operations; thus, a pooling layer is typically used when building neural networks.

The pooling layer is typically follows the convolutional layer, and the fully connected layer generally constitutes the final layers. The fully connected layer is usually a classifier that flattens the result to one dimension by converting it into a single vector that is used as the input of the next layer. The weight of the next FC layer is used to predict the correct label, and the output of the last fully connected layer is the final probability of each label.

Each of the three CNN layer can perform useful calculations; thus, our research combined these three layers to construct a highly accurate CNN model.

2.2. Spin-Orbit Torque MRAM

Spin-orbit torque MRAM (SOT-MRAM) [10] is the generation of MRAM following spin transfer torque MRAM (STT-MRAM). The main difference between STT-MRAM and

SOT-MRAM is that SOT-MRAM uses a more energy-efficient material called spin hall metal (SHM). SHM causes a rotating Hall effect on the application of a write current; this Hall effect creates a spin-torque switch on the magnetic channel of the free layer. SOT-MRAM does not require substantially more write current than STT-MRAM does because the area in which the current flows through SHM is relatively small. SOT-MRAM also has separate read and write paths, which can improve read and write speeds. Thus, we used SOT-MRAM circuit architecture.

An SOT-MRAM cell comprises two word lines, namely the read word line (RWL) and write word line (WWL); two bit lines, namely the read bit line (RBL) and write bit line (WBL); one source line (SL); and two access transistors. The details of SOT-MRAM operation are as follows.

On a rising write signal, the write current from the WBL flows in, and the WWL signal simultaneously activates the access transistor. Thus, the write current can flow through the access transistor. For a written value of 0(1), the current changes from SL(WBL) to WBL(SL). The direction of the free layer's magnetic field can be changed by the spin Hall effect, which is generated by the different current directions.

If the direction of the changed magnetic field is parallel (anti-parallel) to the fixed magnetic field, the effective resistance of the MTJ is R_P (R_{AP}), which has low (high) impedance. By connecting SL to GND and connecting the switch voltage source (V_{write}) to WBL, the direction of the write current can be changed directly.

On a rising read signal, the induced current passes through RBL. The sensing current then passes through the bit cell when the RWL signal switches on the transistor on the RBL side. To read the bit cell, the sense amplifier senses the voltage of the BL. The sensed current and the resistance of the bit cell are known when SL is grounded. Finally, the voltage value of BL can be calculated by the product of the sensing current and the effective resistance (R_P and R_{AP}) of the unit.

2.3. Memory Comparison

Table 1 presents the characteristics of different types of memory for comparison [11]. The read/write speed of MRAM is similar to that of SRAM and DRAM, but the read power consumption is substantially lower than both DRAM and SRAM. In addition, MRAM is nonvolatile memory; thus, MRAM does not consume any energy outside of I/O operations, and data are not lost when power is disconnected. MRAM also has advantages in its manufacture over DRAM and SRAM in that it can be combined with an original digital circuit by adding a masking layer. In the future, MRAM may be able to replace the cache or flash memory in microcomputer units (MCUs). Therefore, MRAM is suitable for the design of CIM circuit architecture.

Table 1. Characteristics of memory architectures for comparison.

Methods	MRAM	SRAM	DRAM	Flash	FeRam
Read Speed	Fast	Fastest	Medium	Fast	Fast
Write Speed	Fast	Fastest	Medium	Low	Medium
Array Efficiency	Med/High	High	High	Med/Low	Medium
Future Scalability	Good	Good	Limited	Limited	Limited
Cell Density	Med/High	Low	High	Medium	Medium
Nonvolatile	Yes	No	No	Yes	Yes
Endurance	Limited	Infinite	Infinite	Limited	Limited
Cell Leakage	Low	Low/High	High	Low	Low
limited	Yes	Yes	Limited	Limited	Limited
Complexity	Medium	Low	Medium	Medium	Medium

2.4. Computing Bit-Wise Logical Operations in SOT-MRAM

2.4.1. Energy Efficient Method of AND/OR Operations in SOT-MRAM

A physics-based compact model for a three terminal PMTJ is proposed in [12], which models the magnetic, electrical, and thermal behaviors of a PMTJ controlled through SOTs. It considers the effects of both damping-like and field-like SOTs on device behavior. Moreover, the model tackles the dynamic behavior of the self-heating process within the device. However, the compact model does not design in-memory computing architecture for a convolutional neural network based on spin-orbit torque MRAM. After that, an energy efficient method of AND/OR operations in SOT-MRAM [8] is proposed, which is the first generation of MRAM using a changing magnetic field to change its internal resistance to store a bit (0 or 1). When a fixed current is input to read 0 or 1, different voltages can be obtained. Thus, if a current is input to two MRAMs, four different voltages are obtained, as presented in Figure 1b. The AND and OR results of these two cells can be obtained and can achieve CIM in the design of sensing amplifiers. However, this method of reading two cells simultaneously has disadvantages. As presented in Table 2, the voltage difference between two cells, namely voltage gap, is approximately 1 mV but can be as low as 0.5 mV. This slight gap is a substantial challenge for designing sense amplifier (SA) and also reduces the robustness of this circuit. Therefore, we adopted the circuit presented in Figure 1a to overcome this problem.

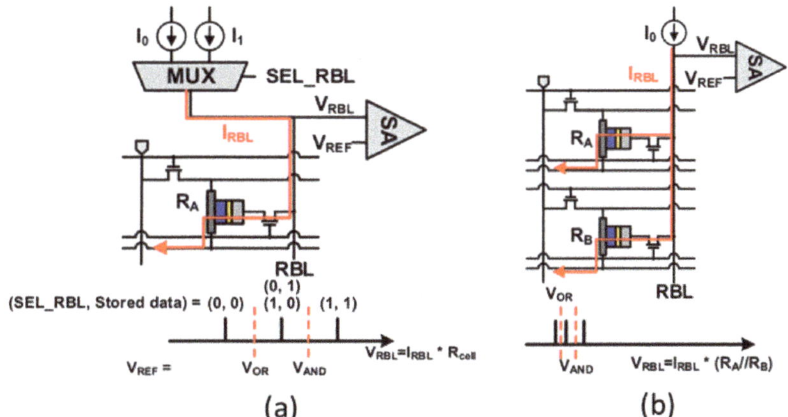

Figure 1. SOT-MRAM circuit for CIM [8]. (**a**) depicts an improved circuit with lower read and write current; the circuit is also more robust when executing CIM operations. (**b**) presents a conventional circuit that requires a larger current for reading and writing when performing CIM operations to achieve the same robustness as the improved circuit.

Table 2. Read results for the circuit in Figure 1b (1: R_{AP} = 6000 (ohm), 0: R_P = 3000, input current 1 µA).

Read Value of Two Cells	Parallel Resistance (ohm)	Read Voltage (mV)	Voltage Gap (mV)
00	1500	1.5	-
01	2000	2	0.5
10	2000	2	1
11	3000	3	1

First, the voltage of the cell is measured. This voltage is used to determine whether the input current is I1 or I0, and the ratio of I1 to I0 is equivalent to the ratio of R_{AP} to R_P. As indicated in Table 3, the voltage difference is larger, and the calculation can, thus, be more robust. Figure 2 reveals that, for the same read current, the voltage gap required by the circuit of Figure 1a is greater than Figure 1b, indicating that the circuit has a greater robustness. Therefore, low current and low power consumption can be used to set the same voltage threshold.

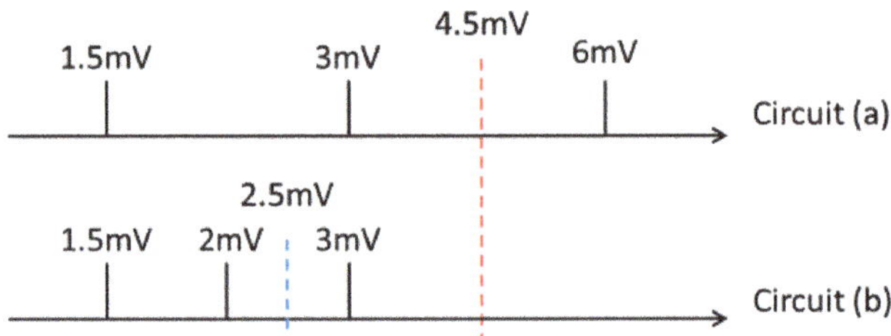

Figure 2. Differences between circuits in (**a**,**b**) during execution and operation.

Table 3. Read result of the circuit of Figure 1a (1: R_{AP} = 6000 (ohm), 0: R_P = 3000 (ohm), I1 = 1 µA, I0 = 0.5 mA).

Reading Value of Two Cells	Reading Voltage (mV)	Voltage Gap (mV)
00	1.5 mV	-
01	3 mV	1.5 mV
10	3 mV	3 mV
11	6 mV	3 mV

2.4.2. Majority Operation

The majority function returns an output of 1 if more than two signals of the three inputs are 1. The truth table of the majority logic operation is presented in Table 4. According to Table 4, the result of majority logic is equivalent to the Cout of a full adder. Therefore, this characteristic can be used to implement a full adder in memory.

Table 4. Truth table of the majority function.

A	B	Cin	Cout
0	0	0	0
0	0	1	0
0	1	0	0
1	0	0	0
0	1	1	1
1	0	1	1
1	1	0	1
1	1	1	1

2.4.3. Majority Decision in Memory

Kirchhoff's circuit law indicates that the input current at a node is equal to the output current at a node; this property can be used to implement a current adder. By matching with a corresponding V_{ref}, the majority result can be obtained, as presented in Figure 3. Therefore, Cout in memory can be quickly obtained after the majority operation is performed.

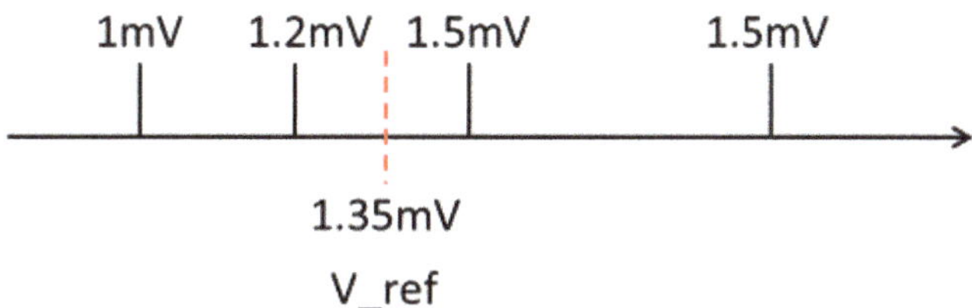

Figure 3. Voltage distribution of majority, where V_ref = 1.35 mV.

2.5. IMCE

IMCE [5] is a method from a paper published by Angizi et al. in 2018 [5]. The method uses bitwise in-memory computing to execute calculations. The method requires 3^2 steps to execute an AND operation for a 3-bit value. For each row, a bit count and shift are required. The total is then summed to complete a 4×4 convolution. In addition, the bitwise in-memory computing requires additional circuits that increase the power consumption of the critical path. The advantage of the method proposed in this paper is that weight and data are stored in the same memory to facilitate calculation. However, this method requires additional circuits and more cycles to complete the calculation in memory. Overall, the power consumption and critical path of ICME are much greater than those of our method.

2.6. Energy-Efficient CIM Architecture

Kim et al. formulated another method in 2019 [8]. Their method first executes an AND operation to obtain partial sums, and it then uses a full adder circuit to complete all steps in sequence. If multiple bit lines are executed together, the final result must be calculated through the outer processing of tempsum1 + tempsum2 $<<$ 1. The advantage of the method is that it can run the full adder in memory without the use of additional circuits, but it requires more cycles to complete. The read and write operations must be executed numerous times, and the control of the method is also complicated. These shortcomings cause the data to be read slowly from the CPU, which is partly because the data are stored in the same destination address, meaning that each address can only read one data.

3. Proposed Architecture

Distributed Arithmetic (DA) was first introduced by Croisier et al. in 1989 [13]. It is an effective method of operations based on memory access and is effectively a bit-serial operation. The execution time depends on the clock speed, read/write speed of the memory, and the length of the operation bit. Figure 4 presents a DA circuit.

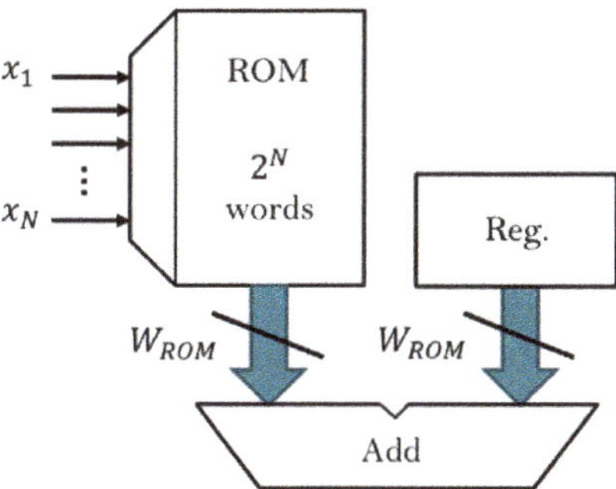

Figure 4. DA circuit.

Let us consider the convolution of two N-point vectors x_i and a fixed coefficient vector h, which is expressed as follows:

$$y = \sum_{i=0}^{N-1} x_i h_i, \tag{3}$$

where $h = [h_0, h_1, h_2 \ldots, h_{(n-1)}]$ and the input vector os $x = [x_0, x_1, x_2 \ldots, x_{(n-1)}]$. Let us assume that x_i is expressed in B-bit two's complement representation as follows.

$$x_i = -x_i + \sum_{j=1}^{B-1} x_{ij} 2^{-j} \tag{4}$$

By substituting (3) in (4), the output y can be expressed in an expanded form as follows.

$$y = \sum_{i=0}^{N-1} x_{i0} h_i + \sum_{j=1}^{B-1} [\sum_{i=0}^{N-1} x_{ij} h_i] 2^{-j} \tag{5}$$

Because h_i is constant, there exist 2^n possibilities for $\sum_{i=0}^{N-1} x_{ij} h_i$ for $j = 1, 2, \ldots, (B-1)$. However, these values can be calculated and stored in memory ahead of time. Thus, we can obtain a partial sum by the bit sequence as the address of the read memory. Therefore, the inner product can be calculated through an accumulation loop of B shifter-adders and by reading the value of the corresponding bit sequence. In our method, DA and the CIM structure are combined to overcome the challenges of the aforementioned model [14].

3.1. Integral Architecture

Figure 5 presents a memory circuit comprising eight banks. Each bank comprises 16 mats, and each mat has four cell arrays of size 16×1024 for a total of 1 megabyte. The control circuit can only control eight banks simultaneously. Each operation can run 16 parallel mats operations, and each mat has 4 cell arrays. In addition, each cell array can perform four 3×3 convolutions simultaneously; thus, the memory architecture can execute 64 convolutions in parallel.

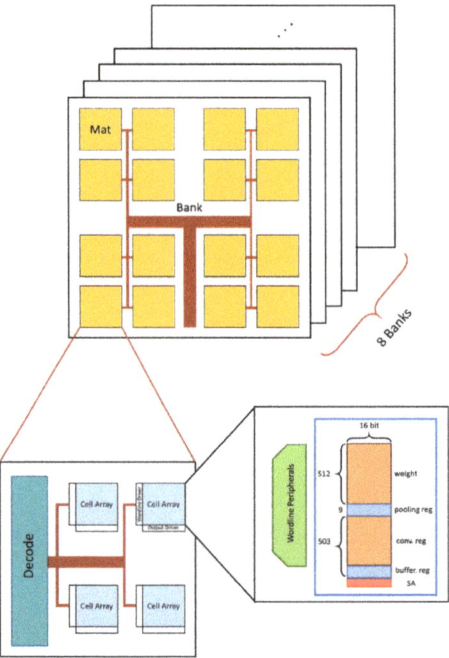

Figure 5. Circuit architecture of memory.

3.2. Achievements Made by the New Architecture

Figure 6 presents our proposed CIM circuit architecture integrating a DA circuit architecture in memory without any digital circuits, such as a full adder or shifter circuit, to implement a DA calculation algorithm. In addition, the CIM architecture requires no additional weighting data; correspondingly, placing the results data and the buffer register on the same cell array can reduce both data access time and power consumption. The execution speed depends on the clock frequency, read speed of memory, and length of the calculation unit. Therefore, the novel CIM architecture performs faster than the traditional DA architecture and has lower power consumption because of the operations performed in memory.

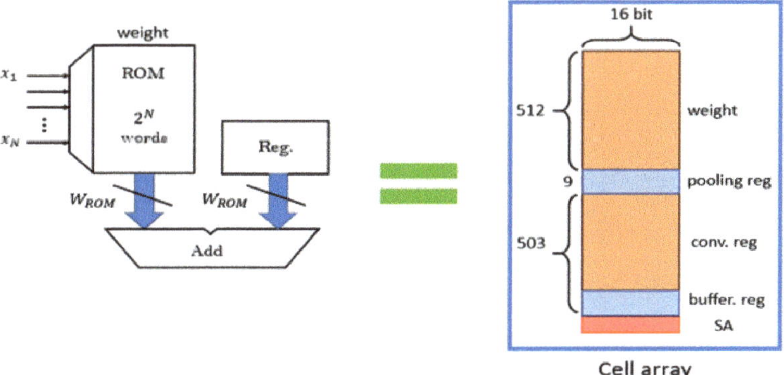

Figure 6. Our CIM circuit architecture.

Due to the advantages of the DA algorithm, the precalculated partial sum can be stored in the memory, and the shifter adder can then be used to accumulate the sum of each part. Therefore, our approach uses only a shifter adder and does not require a multiplier with a long critical path or a large area. Our new CIM architecture avoids the lengthy execution steps and additional circuits required by previous methods.

The main components of the DA architecture circuit are the read-only memory (ROM), reg buffer, full adder, and shifter. The following section describes the structure, operation, and implementation of these components to achieve the DA architecture in memory.

3.2.1. Build ROM and Register (Reg) Buffer in the Memory

MRAM is nonvolatile memory, and its read speed is similar to that of DRAM; thus, MRAM is suitable as the storage unit for a DA architecture. To increase the efficiency of in-memory calculation execution and to achieve lower latency and read/write power consumption, the weighted memory and buffer register stored in the CIM are placed on the same cell array shown, as presented in Figure 7. In addition, these defined memory sizes can be changed because the entire memory space, not only one specific part of the memory, can complete CIM operations.

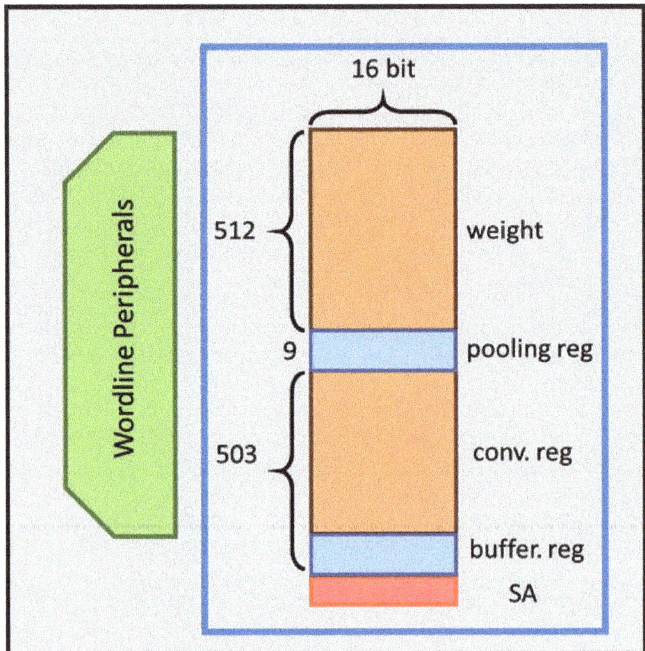

Figure 7. Storage unit configuration.

3.2.2. Shifter

Because the shifter is unavailable in traditional memory, our method adds N-Metal-Oxide Semiconductor (NMOS) and P-Metal-Oxide Semiconductor (PMOS) to the SA circuit architecture, as presented in Figure 8. This change enables the output of the SA to be written into different columns based on shifter control without reading data out of the cell array or rewriting; these processes would otherwise extend the read/write time.

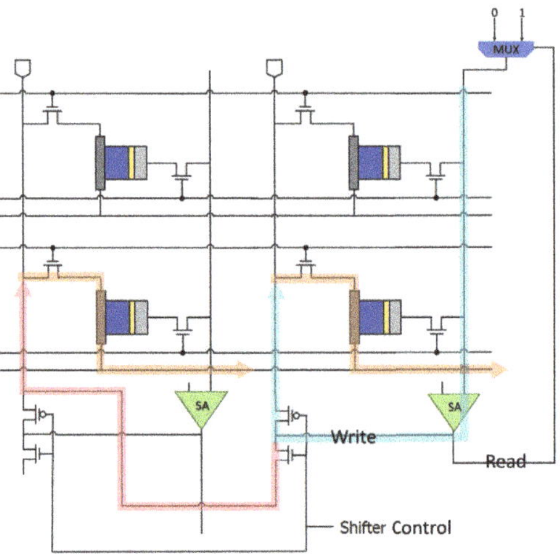

Figure 8. Shift circuit.

3.2.3. Shifter Full-Adder

This unit is used to complete a full adder operation. First, it calculates MAJ(A,B,Cin) to obtain Cout and obtain a sum in parallel in the following step. Then, $(A \oplus Cin) \oplus B$

is performed to obtain sum-reg. Finally, the left shift to the sum is executed for the next shift-adder operation, as presented in Figure 9.

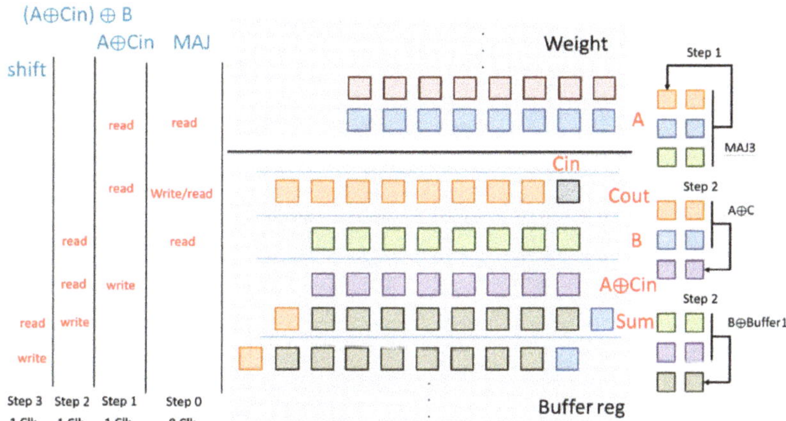

Figure 9. Steps of the shift adder.

4. Experimental Process and Result

4.1. Experimental Process

The simulation was divided into three stages, as presented in Figure 10. First, the Hspice tool was used to obtain a circuit level result. Second, the result was sent to the Nvsim model for simulation by using the memory architecture-level data obtained in stage 1. Subsequently, read/write power consumption and the memory delay were sent to GEM5 to execute system-level calculations. After these three stages of simulation, the written LeNet-5 algorithm could be executed in GEM5 to obtain the read/write power consumption of the entire algorithm. In our simulation, the parameters are set as follows. We considered the setup of our SPICE and process file, referring to references [15,16], respectively, with read voltage = 6 mV, current = 1 uA, and register = 6 kΩ. For the setup of NVSIM, we choose 1 MB memory with 8 banks, in which each bank has 16 MATs; each MAT has 4 cell arrays; and each cell array is 16×1024 bits in size. Additionally, we considered accelerators using 16-bit gradients; we select MNIST as our benchmark dataset, along with the LeNet-5 NN architecture. For the CNN layers of each 32×32 image, we developed a bitwise CNN with six convolutional layers, two average pooling layers and two FC layers, which are equivalently implemented by convolutional layers. After collecting this information, we could conduct comparisons using the experimental data. The detailed process of each stage is described in the following paragraph.

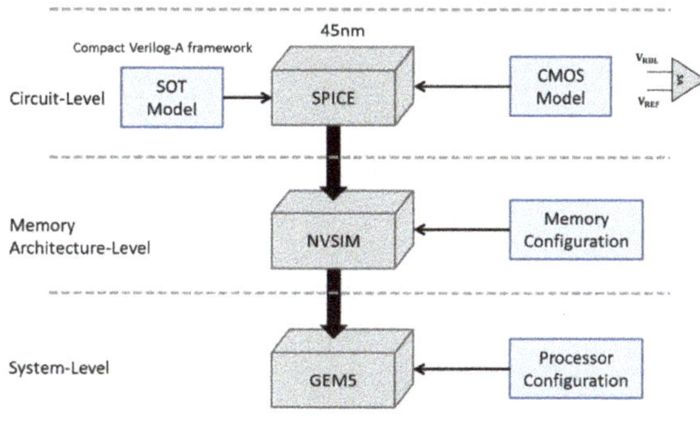

Figure 10. Experimental architecture.

4.1.1. SOT-MRAM Simulation

The SOT-MRAM model is built at the circuit level. The MRAM model used in our research is the same as that presented in [17]. That study's authors provided the SOT-MRAM Verilog-A model file that is used to facilitate simulations and verifications of the real-world performance of MRAM memory. Figure 11 presents the simulation results of MRAM Verilog-A in Cadence Virtuoso.

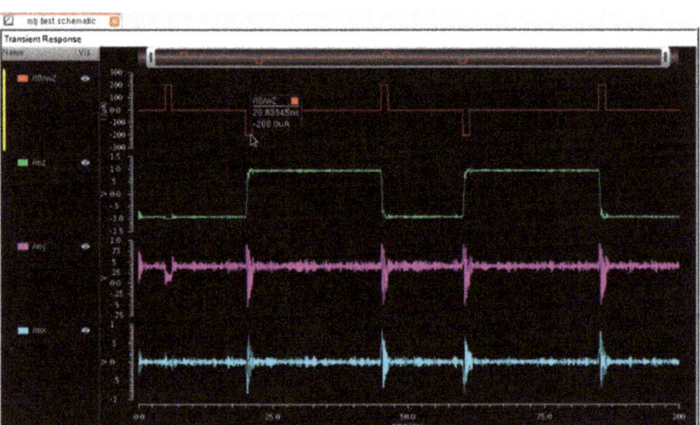

Figure 11. Simulated waveform of an MRAM cell.

4.1.2. Processand Sensing Amplifier (SA)

We used NCSU FreePDK 45 nm [16] to simulate the SA circuit architecture and the digital circuit synthesis in Hspice. In addition, we chose StrongARM Latch [18], which consumes zero static power and has low latency. Thus, it is suitable for edge computing in the CIM architecture presented as Figure 12.

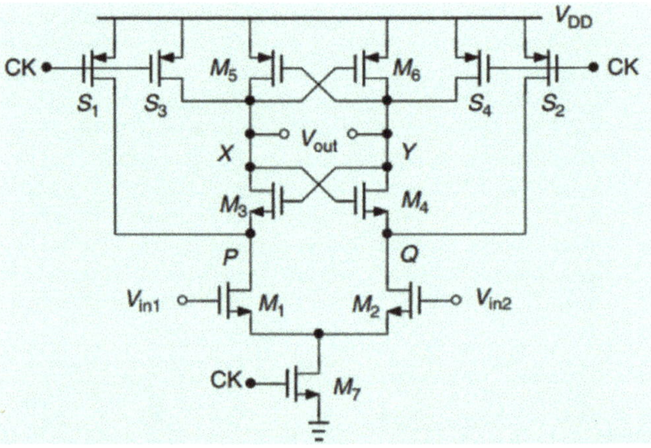

Figure 12. StrongARM Latch.

4.1.3. Nvsim

Nvsim [19] is a circuit-level model used to estimate the performance, energy, and area of new nonvolatile memory (NVM). Nvsim supports various NVM technologies, including STT-MRAM, PCRAM, ReRAM, and traditional NAND flash; thus, it was used and was modified to match the architectures we chose for simulation.

4.1.4. Gem5

Gem5 [20] is a modular discrete event-driven simulator for a full-system that combines the advantages of M5 and GEMS. M5 is a highly configurable simulation framework that supports a variety of ISAs and CPU models. In addition, GEMS complements the features of M5 by providing a detailed and flexible memory system, including multiple cache consistency protocols and interconnection models. It is a highly configurable architecture simulator that integrates multiple ISAs and multiple CPU models. In our experiment, we used a single-core Arm A9 CPU clocked at 2 Ghz as the CIM CPU for simulation analysis. Figure 13 presents the entire simulation process in Gem5. First, C code was compiled into a binary file, and Gem5 was then used to simulate the binary file and obtain a *states.txt* file containing data on the simulated CPU cycles and the read/write times of memory. Then, CPU power consumption could be obtained with the Mcpat tool.

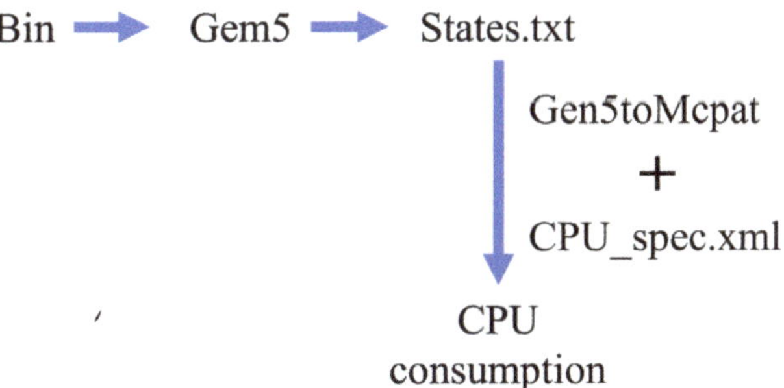

Figure 13. Simulation of an A9 processor running with LeNet-5.

4.2. Experimental Result

4.2.1. IMCE vs. Our Method

We analyzed the number of access times in reading, writing, and overall reading/writing separately; the results were then compared with those in previous IMCE studies. Figure 14 presents the convolution algorithm used for this comparison. The input image data and weight were both 8-bit. Figure 15 present three comparisons of reading and writing times. For reading times only, as presented in Figure 15a, we observed that our method is 49.9% faster that IMCE. As presented in Figure 15b, our method was 22.7% faster than IMCE in writing times. Finally, with regard to overall reading/writing times, as indicated in Figure 15c, our method was 43.3% faster than IMCE overall. This improvement was due to the use of the DA algorithm to substantially reduce the number of reads and, thus, reduce the power consumption during CIM by replacing multiplication with a lookup table.

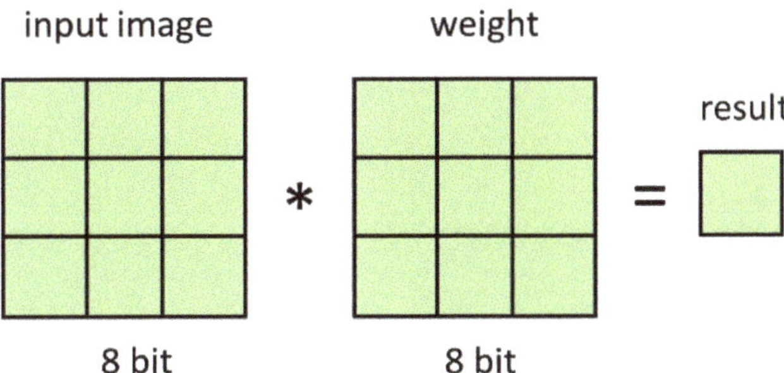

Figure 14. Convolution process.

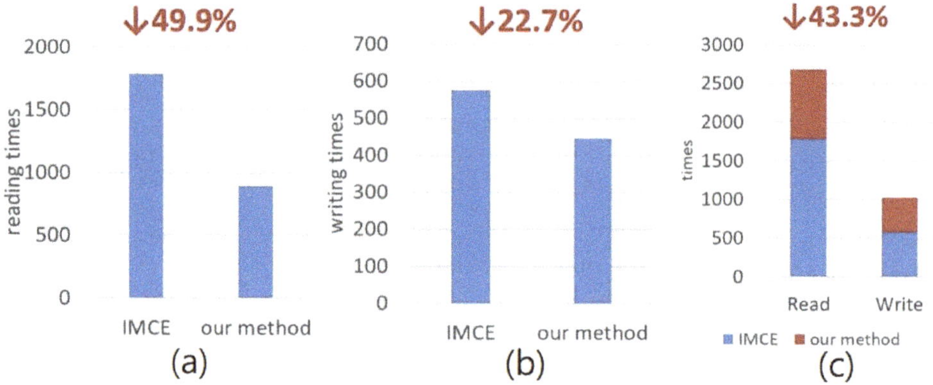

Figure 15. (**a**) Reading times, (**b**) writing times, and (**c**) comparison.

4.2.2. Traditional Non-CIM Architecture and CIM Architecture

To determine whether the CIM architecture can effectively reduce computing power consumption, we used a non-CIM architecture and the CIM architecture to run the same NN as presented in Figure 16; this algorithm was also used in our experimental analysis. The main focus of our research is the ConV operation; therefore, the FC layer in the NN handled the CPU operation, and the CIM circuit architecture handled the ConV operation. Table 5 presents a comparison of the power consumption of the two architectures. Figure 17a shows the comparison of CPU power consumption. The difference is primarily because the convolution operation consumes the most power. In the CIM architecture, the convolution operation is moved to memory. Thus, the CPU does not perform the convolution operation, greatly reducing power consumption. Figure 17b illustrates the memory power consumption for comparison. CIM architecture can minimize the data transmission path and more greatly reduce the total memory power consumption. Moreover, because convolution has been moved to memory and also further reduced power consumption, the total read/write power consumption of the memory was lower. Figure 17c presents a comparison of the total power consumption, revealing that the CIM circuit architecture has lower overall power consumption.

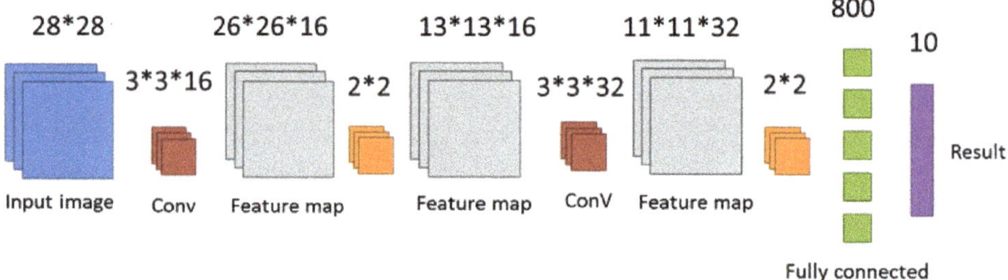

Figure 16. LeNet-5 NN architecture.

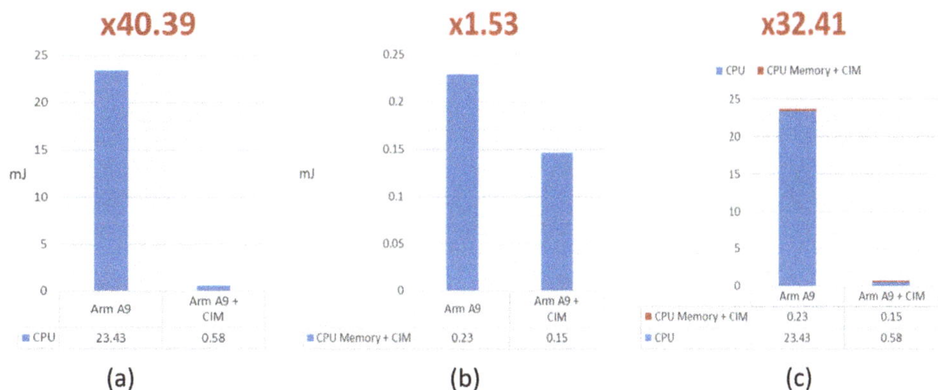

Figure 17. (a) CPU, (b) memory, and (c) overall.

Table 5. Comparison of non-CIM architecture and CIM architecture.

-	Arm A9	Arm A9 + CIM
CPU	23.43	0.5837034
CPU Memory	0.23	0.0006997
CIM	0	0.1459011
Total Power	23.66	0.7303042

4.3. Discussion

Our CIM architecture can be used for CPUs, GPUs, FPGAs, and ASIC in different design manners. In-memory computing has two advantages: making computing faster and scaling it to potentially support petabytes of in-memory data. In-memory computing utilizes two key technologies: random-access memory storage and parallelization. When the CPU/GPU processes data from the main memory, frequently used data are stored in fast, energy-efficient caches to enhance performance and energy efficiency. However, in applications that process large amounts of data, most data are read from the main memory because the data to be processed are very large compared to the size of the cache. In this case, the bandwidth of the memory channel between the CPU/GPU and the main memory becomes a performance bottleneck, and a lot of energy is consumed to transfer data between the CPU/GPU and the main memory. To alleviate this bottleneck, the channel bandwidth between the CPU/GPU and main memory needs to be extended but if the current CPU/GPU's number of pins has reached its limit, further bandwidth improvement faces technical difficulties. In a modern computer structure where data storage and data calculation are separated, such a memory wall problem will be inevitably raised. Our CIM architecture is used to overcome the aforementioned bottleneck by performing operations in memory without moving data to the CPU/GPU. Additionally, our CIM architecture can also be implemented in FPGA or as an ASIC design under the assumption that the MRAM can be well taped out.

5. Conclusions

We proposed a new SOT-MRAM-based CIM architecture for a CNN model that can reduce both power consumption and read/write in comparison with conventional CNN CIM architectures. In addition, our method does not require additional digital circuits, enabling the MRAM cell to retain the advantages of memory for data storage. By conducting a series of experiments, compared with the ICME method [5], our proposed method reduces read times by 49.9%, write times by 22.7%, and overall read/write times by 43.3%.

Additionally, we evaluated that a CIM model running on an Arm A9 CPU can significantly reduce power consumption. In this paper, we did not tackle the changing magnetic field of MRAM. We used highly configurable architecture simulators SPICE, NVSIM, and GEM5 models to evaluate our proposed SOT CIM-based architecture. Quantifying the changing/switching magnetic field of the MRAM is an open issue. In the future, we will collaborate with industries to realize it.

Author Contributions: Conceptualization, J.-Y.H. and Y.-T.T.; methodology, J.-Y.H. and Y.-T.T.; validation, J.-Y.H., J.-L.S., Y.-T.T., S.-Y.K. and C.-R.C.; formal analysis, J.-Y.H. and Y.-T.T.; investigation, J.-Y.H., J.-L.S., Y.-T.T. and C.-R.C.; resources, Y.-T.T., S.-Y.K. and C.-R.C.; data curation, J.-Y.H. and Y.-T.T.; writing—original draft preparation, J.-Y.H. and Y.-T.T.; writing—review and editing, Y.-T.T., S.-Y.K. and C.-R.C.; supervision, Y.-T.T., S.-Y.K. and C.-R.C.; project administration, Y.-T.T. and C.-R.C.; funding acquisition, Y.-T.T. and C.-R.C. All authors have read and agreed to the published version of the manuscript.

Funding: This work is supported by the Ministry of Science and Technology, Taiwan, under grant MOST 109-2923-E-035-001-MY3, MOST 110-2112-M-033-013, and MOST 110-2221-E-035-034-MY3.

Conflicts of Interest: The authors declare no conflict of interest.

References

1. Ou, Q.-F.; Xiong, B.-S.; Yu, L.; Wen, J.; Wang, L.; Tong, Y. In-Memory Logic Operations and Neuromorphic Computing in Non-Volatile Random Access Memory. *Materials* **2020**, *13*, 3532. [CrossRef] [PubMed]
2. Zou, X.; Xu, S.; Chen, X.; Yan, L.; Han, Y. Breaking the von Neumann Bottleneck: Architecture-Level Processing-in-Memory Technology. *Sci. China Inf. Sci.* **2021**, *64*, 1–10. [CrossRef]
3. He, K.; Zhang, X.; Ren, S.; Sun, J. Deep Residual Learning for Image Recognition. In Proceedings of the IEEE Conference on Computer Vision and Pattern Recognition, Honolulu, HI, USA, 21–26 July 2016; pp. 770–778.
4. Deng, Q.; Jiang, L.; Zhang, Y.; Zhang, M.; Yang, J. DRACC: A Dram based Accelerator for Accurate CNN Inference. In Proceedings of the 55th Annual Design Automation Conference, San Francisco, CA, USA, 24–29 June 2018; pp. 1–6.
5. Angizi, S.; He, Z.; Parveen, F.; Fan, D. IMCE: Energy-efficient Bitwise In-emory Convolution Engine for Deep Neural Network. In Proceedings of the 23rd Asia and South Pacific Design Automation Conference, Jeju Island, Korea, 22 January 2018; pp. 111–116.
6. Chi, P.; Li, S.; Xu, C.; Zhang, T.; Zhao, J.; Liu, Y.; Wang, Y.; Xie, Y. Prime: A Novel Processing-in-Memory Architecture for Neural Network Computation in ReRAM-based Main Memory. *ACM SIGARCH Comput. Archit. News* **2016**, *44*, 27–39. [CrossRef]
7. Li, S.; Niu, D.; Malladi, K.T.; Zheng, H.; Brennan, B.; Xie, Y. DRISA: A Dram-based Reconfigurable In-situ Accelerator. In Proceedings of the 50th Annual IEEE/ACM International Symposium on Microarchitecture, Boston, MA, USA, 14–17 October 2017; pp. 288–301.
8. Kim, K.; Shin, H.; Sim, J.; Kang, M.; Kim, L.-S. An Energy-Efficient Processing-in-Memory Architecture for Long Short Term Memory in Spin Orbit Torque MRAM. In Proceedings of the International Conference on Computer-Aided Design, Westminster, CO, USA, 4–7 November 2019; pp. 1–8.
9. Albawi, S.; Mohammed, T.A.; Al-Zawi, S. Understanding of a Convolutional Neural Network. In Proceedings of the International Conference on Engineering and Technology, Antalya, Turkey, 21–23 August 2017; pp. 1–6.
10. Zhang, Y.; Wang, G.; Zheng, Z.; Sirakoulis, G. Time-Domain Computing in Memory Using Spintronics for Energy-Efficient Convolutional Neural Network. *IEEE Trans. Circuits Syst.* **2021**, *68*, 1193–1205. [CrossRef]
11. Xu, T.; Leppänen, V. Analysing Emerging Memory Technologies for Big Data and Signal Processing Applications. In Proceedings of the Fifth International Conference on Digital Information Processing and Communications, Sierre, Switzerland, 7–9 October 2015; pp. 104–109.
12. Kazemi, M.; Rowlands, G.E.; Ipek, E.; Buhrman, R.A.; Friedman, E.G. Compact Model for Spin–Orbit Magnetic Tunnel Junctions. *IEEE Trans. Electron Devices* **2016**, *63*, 848–855. [CrossRef]
13. White, S.A. Applications of Distributed Arithmetic to Digital Signal Processing: A Tutorial Review. *IEEE Assp Mag.* **1989**, *6*, 4–19. [CrossRef]
14. Chen, J.; Zhao, W.; Ha, Y. Area-Efficient Distributed Arithmetic Optimization via Heuristic Decomposition and In-Memory Computing. In Proceedings of the 13th International Conference on ASIC, Chongqing, China, 29 October–1 November 2019; pp. 1–4.
15. Kim, J.; Chen, A.; Behin-Aein, B.; Kumar, S.; Wang, J.P.; Kim, C.H. A Technology-Agnostic MTJ SPICE Model with User-Defined Dimensions for STT-MRAM Scalability Studies. In Proceedings of the 2015 IEEE Custom Integrated Circuits Conference (CICC), San Jose, CA, USA, 28–30 September 2015; pp. 1–17.
16. Ncsu Eda freepdk45. FreePDK45:Contents. 2011. Available online: http://www.eda.ncsu.edu/wiki/ (accessed on 21 December 2020).

17. Alwani, M.; Chen, H.; Ferdman, M.; Milder, P. Fused-layer cnn accelerators. In Proceedings of the 2016 49th Annual IEEE/ACM International Symposium on Microarchitecture, Taipei, Taiwan, 15–19 October 2016; pp. 1–12.
18. Razavi, B. The StrongARM Latch [A Circuit for All Seasons]. *IEEE Solidstate Circuits Mag.* **2015**, *7*, 12–17. [CrossRef]
19. Dong, X.; Xu, C.; Xie, Y.; Jouppi, N.P. Nvsim: A Circuit-Level Performance, Energy, and Area Model for Emerging Nonvolatile Memory. *IEEE Trans. Comput.-Aided Des. Integr. Circuits Syst.* **2012**, *31*, 994–1007. [CrossRef]
20. Binkert, N.; Beckmann, B.; Black, G.; Reinhardt, S.K.; Saidi, A.; Basu, A.; Hestness, J.; Hower, D.R.; Krishna, T.; Sardashti, S.; et al. The Gem5 Simulator. *Acm Sigarch Comput. Archit. News* **2011**, *39*, 1–7. [CrossRef]

Article

Development of a Face Prediction System for Missing Children in a Smart City Safety Network

Ding-Chau Wang [1], Zhi-Jing Tsai [2], Chao-Chun Chen [2] and Gwo-Jiun Horng [3,*]

[1] Department of Information Management, Southern Taiwan University of Science and Technology, Tainan 71005, Taiwan; dcwang@stust.edu.tw
[2] Institute of Manufacturing Information and Systems, National Cheng Kung University, Tainan 70101, Taiwan; p96091115@gs.ncku.edu.tw (Z.-J.T.); chaochun@mail.ncku.edu.tw (C.-C.C.)
[3] Department of Computer Science and Information Engineering, Southern Taiwan University of Science and Technology, Tainan 71005, Taiwan
* Correspondence: grojium@stust.edu.tw

Abstract: Cases of missing children not being found are rare, but they continue to occur. If the child is not found immediately, the parents may not be able to identify the child's appearance because they have not seen their child for a long time. Therefore, our purpose is to predict children's faces when they grow up and help parents search for missing children. DNA paternity testing is the most accurate way to detect whether two people have a blood relation. However, DNA paternity testing for every unidentified child would be costly. Therefore, we propose the development of the Face Prediction System for Missing Children in a Smart City Safety Network. It can predict the faces of missing children at their current age, and parents can quickly confirm the possibility of blood relations with any unidentified child. The advantage is that it can eliminate incorrect matches and narrow down the search at a low cost. Our system combines StyleGAN2 and FaceNet methods to achieve prediction. StyleGAN2 is used to style mix two face images. FaceNet is used to compare the similarity of two face images. Experiments show that the similarity between predicted and expected results is more than 75%. This means that the system can well predict children's faces when they grow up. Our system has more natural and higher similarity comparison results than Conditional Adversarial Autoencoder (CAAE), High Resolution Face Age Editing (HRFAE) and Identity-Preserved Conditional Generative Adversarial Networks (IPCGAN).

Keywords: face aging; generative adversarial network; StyleGAN2; FaceNet; missing child

Citation: Wang, D.-C.; Tsai, Z.-J.; Chen, C.-C.; Horng, G.-J. Development of a Face Prediction System for Missing Children in a Smart City Safety Network. *Electronics* **2022**, *11*, 1440. https://doi.org/10.3390/electronics11091440

Academic Editor: Byung-Gyu Kim

Received: 16 February 2022
Accepted: 25 April 2022
Published: 29 April 2022

Publisher's Note: MDPI stays neutral with regard to jurisdictional claims in published maps and institutional affiliations.

1. Introduction

We will divide this paper into three subsections to describe the introduction: "Status of Missing Children's Cases", "Problems of Current Face Aging Methods", and "Contribution".

1.1. Status of Missing Children's Cases

According to the Federal Bureau of Investigation's National Crime Information Center (NCIC) Missing Person and Unidentified Person Statistics, there will be 365,348 children missing in the United States in 2020 [1]. According to the National Crime Agency's Missing Persons Statistics, it is estimated that over 65,800 children will go missing between 2019 and 2020 [2].

From the above statistical results, it is known that there has always been a situation of missing children. Therefore, we built the Face Prediction System for Missing Children to predict the face of children when they grow up and help parents or police search, detect, and identify missing children. Moreover, we applied our system to the Smart City Safety Network. According to S. P. Mohanty et al.'s description [3], a smart city includes smart infrastructure, smart transportation, smart energy, smart health care, and smart technology. The key to transforming traditional cities into smart cities is information and communication technology (ICT). Smart cities use ICT to solve a variety of urban problems. In addition, M. Lacinák et al. emphasized the importance of safe cities [4,5]. They describe that every smart city must also be a safe city, and a safe city should be regarded as a part of a smart city. A safe city system should include the following features: smart safety systems for surveillance, search, detection and identification, etc. The purpose of our system conforms to the concept of smart city safety, and then we use the concepts of IoE and AIoT to implement our system and form a network. More details can be found in Section 3.

To predict future faces currently, face aging image generation methods in the field of machine learning can be used. However, the existing face aging model only considers the facial features that only older people have. In fact, head size and genetics can also affect appearance. Therefore, current face aging methods cannot well predict children's faces when they grow up. Section 1.2 describes the problems of current face aging methods in detail.

1.2. Problems of Current Face-Aging Methods

We refer to many face-aging image generation methods, such as CAAE [6] extended by VAE, F-GAN [7], HRFAE [8] and IPCGAN [9] extended by GANs. Details can be found in Section 2.3. Overall, these face aging methods add or smooth some irregular wrinkles on the face, making the generated results appear older or younger. However, these methods do not work for children under 12; they only work for adults. According to medical research [3–10], the period from 0 years old to adolescence is the fastest-growing period for human beings. The appearance (including facial appearance, body shape, etc.) will change greatly. Therefore, if we only consider the facial features that only older people have and do not consider other factors (such as head size and genetics, etc.), it is impossible to predict children's faces when they grow up.

Figure 1 illustrates the problems of current face-aging methods. In Figure 1, (a) is the original image, whose age is between 0 and 20 years old, while (b)~(d) is the face image converted by using the original image through the Group-GAN model; (b) is between 20 and 40 years old, (c) is between 40 and 60 years old, and (d) is over 60 years old.

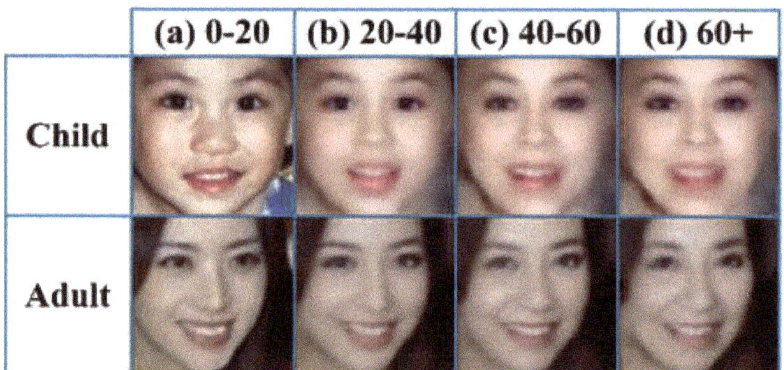

Figure 1. Group-GAN face aging results.

We can see from Figure 1 that the child and the adult are transformed from (a) to (b)~(d), respectively. Two problems arise during this transition:

1. The difference between the child and adult transition from (a) to (b) is small, especially when the child (b) does not look like they are between 20 and 40 years old. The reason is that most aging models only consider the facial texture and do not consider that the head's size will also change with age;

2. The result of a child's transition from (a) to (d) is very unnatural, while the transition of an adult from (a) to (d) is relatively natural. The reason is not only because the size of the head is not taken into account, but also because people grow most rapidly before puberty (children under the age of 12), so the appearance changes very greatly. Therefore, it is not enough to only consider the facial texture for face prediction.

Overall, because existing aging models only consider the appearance characteristics that older adults have, they cannot predict children's future faces.

We propose a Face Prediction System for Missing Children. In addition to taking into account the facial features of children before their disappearance, it also considers the facial features of their blood relatives. According to genetics [10–13], life on earth mainly uses DNA in the blood as genetic material so that the offspring will have the parent's traits (traits including appearance and disease, etc.). In addition, according to Mendelian inheritance [14,15], human looks are mainly determined by genetics, which means that children are born with traits such as appearance and diseases that are part of their parents. Our system takes into account human genetics and then estimates and predicts children's future faces. It is a more reasonable estimate than other aging models. No one has proposed this prediction method to date.

1.3. Contribution

Suppose a parent wants to search, detect and identify their child from a group of unidentified children. If every unidentified child had a DNA paternity test, there would be a lot of cost and time waiting for the test. If parents used our system, they could quickly confirm the possibility of blood relation with any unidentified child. Additionally, there are the following benefits:

- We can directly eliminate this pairing when our system's face prediction image and any child's face image are low in similarity. This is useful for narrowing down the search;
- When our system's face prediction image and any child's face image are high in similarity, we can use this pairing and then conduct the DNA paternity test. It is faster and less expensive than DNA paternity testing for every unidentified child.

Overall, we have the following two contributions:

1. Our system takes into account the facial features of children's blood relatives, and the output predictions are approximately 75% more similar to the expected results. When parents search for missing children, our system helps to eliminate low similarity matches and narrow the search;
2. Parents can quickly and inexpensively confirm the possibility of blood relation with any child.

2. Related Work

2.1. Generative Adversarial Networks (GANs)

GANs [15–17] are unsupervised learning networks trained only through images without labels. A GAN is mainly composed of a generator and discriminator network. The final goal of a GAN is for the generator to randomly create real-looking images that cannot be distinguished from training images. Many scholars have developed different methods and applications based on the concept of GANs. For example, in the field of image generation, the Progressive GAN [18] proposed by NVIDIA Tero Karras et al. can randomly generate high-resolution images. There is also StyleGAN [19] proposed in 2019, which follows the training network of Progressive GAN and has the function of style conversion, which can control the changes of different styles of images.

However, due to the droplet and phase artifact problems in StyleGAN, StyleGAN2 [20] was proposed in 2020 to solve the above problems and make the output results more natural. StyleGAN has had a huge impact on image generation and editing, and many scholars have used StyleGAN for different studies. For example, Image2StyleGAN [21,22], SEAN [23], Editing in Style [24], StyleFlow [25], Pixel2style2pixel [26], StyleCLIP [27] and StyleMapGAN [28] are all image generation and editing methods developed based on StyleGAN.

Figure 2 is the generator architecture of StyleGAN [18–20]. The generator of StyleGAN consists of the mapping network and the synthesis network. The mapping network is a non-linear network using an 8-layer MLP. Its input is the latent code (or latent variable) z in latent space Z, and the output is the intermediate latent code (or dlatents) w in intermediate latent space W. The latent space is simply a representation of compressed data in which similar kinds of data points will be closer in the latent space. Latent space is useful for learning data features and finding simpler data representations for analysis. We can interpolate data in the latent space and use our model's decoder to 'generate' data samples. The purpose of the mapping network is to convert the input z to w. Because the use of z to control image features is limited, a mapping network is needed to convert z to w, which is used to reduce the correlation between features and to control the generation of different images.

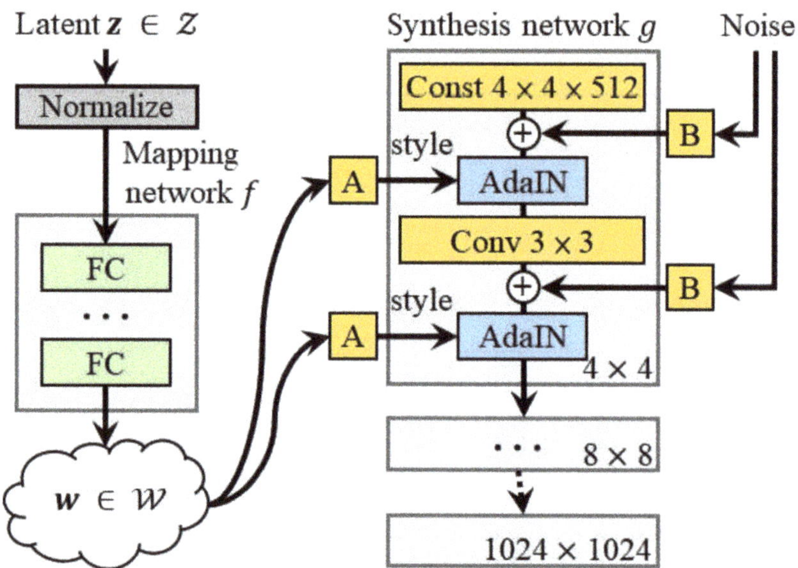

Figure 2. Generator architecture of StyleGAN [18–20].

The synthesis network is used to generate images of different styles and add affine transformation A and random noise B to each sub-network layer. A is used to control the style of the generated image, which can affect the pose of a face, identity features, etc. B is used for the details of the generated image and can affect details such as hair strands, wrinkles, skin color, etc.

Figure 3 shows the style mixing result of StyleGAN. Sources A and B are pre-trained models using StyleGAN to project the images into the corresponding latent space. Finally, the images are directly generated by the latent code. The coarse styles from source B mainly control the coarser low-resolution features (no more than 8 × 8), affecting posture, general hairstyle, facial shape, etc. The middle styles from source B mainly control the finer features of the middle resolution (16 × 16 to 32 × 32), including facial features, hairstyles, opening or closing of eyes, etc. The fine styles from source B mainly control the relatively high-quality, high-resolution features (64 × 64 to 1024 × 1024), affecting the color of eyes, hair, and skin.

Figure 3. Examples that were generated by a mixture of Source A and B latent codes using Style-GAN [19,20].

Table 1 shows the training results of StyleGAN and StyleGAN2. The ↑ indicates that higher is better and ↓ that lower is better. In 2020, StyleGAN2 improved the shortcomings of SyleGAN, including the droplet and phase artifact problem, and enhanced the quality of SyleGAN, including weight demodulation, path length regularization and no progressive growth. Because StyleGAN2 can generate higher quality images and train faster, this study mainly uses the pre-trained model of StyleGAN2 for style mixing and image projection processing.

Table 1. Training results of StyleGAN [19] and StyleGAN2 [20].

	Configuration	FFHQ, 1024 × 1024			
		FID ↓	Path Length ↓	Precision ↑	Recall ↑
A	Baseline StyleGAN [19]	4.40	212.1	0.721	0.399
B	+ Weight demodulation	4.39	175.4	0.702	0.425
C	+ Lazy regularization	4.38	158.0	0.719	0.427
D	+ Path length regularization	4.34	122.5	0.715	0.418
E	+ No growing, new G & D arch.	3.31	124.5	0.705	0.449
F	+ Large networks (StyleGAN2 [20])	2.84	145.0	0.689	0.492

The " + " in the table represents the experimental results based on StyleGAN (A) plus (B) to (F) configurations.

2.2. FaceNet

FaceNet [29] is a unified framework for solving recognition and verification problems proposed by Google. According to Florian Schroff et al., FaceNet mainly uses convolutional neural networks to analyze face information and project the information into Euclidean space. The similarity between the two can be directly calculated by calculating the distance in the space. In 2015, FaceNet received the highest accuracy score of 99.63% in LFW [30] and received attention for this. At present, the development of face recognition and verification is quite mature, including OpenFace [31], Deep-Face [32], VGG-Face [33,34], DeepID [35–38], ArcFace [39] and Dlib [40], all of which have over 90% accuracy. The highest accuracy face-recognition model today is VarGFaceNet [41]. We used Euclidean distance (L2) to calculate distance and converted it into similarity.

2.3. Image Generation Method of Face Aging

In the field of image generation, variational autoencoder (VAE) [42] and generative adversarial network (GAN) [17] methods are the mainstream.

2.3.1. VAE

Research methods extended by VAE include the adversarial autoencoder (AAE) [43,44], the conditional adversarial autoencoder (CAAE) [6], and the conditional adversarial consistent identity autoencoder (CACIAE) [45], etc. AAE is a training method that combines the encoder-decoder idea of VAE and the generator-discriminator of GAN. CAAE is a face-aging method proposed by Z. Zhang et al. It builds a discriminator based on AAE to make the generated images more realistic. CAAE can learn the face manifold and achieve smooth age progression and regression so that the results can appear more aged or younger. In addition, the CACIAE proposed by Bian et al. can reduce the loss of identity information, making the results more realistic and age-appropriate. In the experimental results, our system is compared with CAAE. Since CAAE only considers facial lines, it cannot predict the appearance of children when they grow up.

2.3.2. GANs

The methods of synthesizing face images using GANs can be divided into two categories: translation-based and condition-based.

Translation-Based Method

The translation-based face image synthesis method converts any set of style images into another set of style images. This concept first came from Cycle-GAN [46], proposed by Zhu et al. Its advantage is that it does not require the pairing of two collection domains, making it available for face-style transfer, unlike pix2pix [47], which must have two or two. Only paired data can be used for training. The disadvantage is that it can only be converted between two domains, so later, Choi et al. proposed StarGAN [48], which can learn multiple domains and solve Cycle-GAN's problem.

In terms of face-aging models, Palsson et al. proposed F-GAN [7], based on the style transfer architecture developed by Cycle-GAN. F-GAN combines the advantages of Group-GAN and FA-GAN. When the age span is large (about 20 years old or more), because the effect is better, Group-GAN is used for face conversion, and FA-GAN is used on the contrary. The problem with F-GAN is that it cannot be converted naturally, and the image quality is low. After 2018, because StyleGAN provides an FFHQ dataset, it became easier to generate high-quality images, but it has artifact problems. Subsequently, Shen et al. proposed InterFaceGAN [49], which can semantically edit the learned latent semantic information (for example, changing age, gender and angle, etc.) and repair the artifacts in the image, making the resulting image more natural. Although it produces higher-quality images, it is not suitable for predicting the appearance of children because it only takes into account the texture of the face.

Condition-Based Method

The condition-based face image synthesis method can be regarded as a supervised GAN. It adds an additional condition to the generator's input and the discriminator. The condition can be a label or a picture, etc. The function guides the generator and the discriminator towards training on this condition. This concept first came from cGAN [50–52], proposed by Mirza et al. It has a better effect than the original GAN, so it has been widely used in the future.

In terms of face-aging models, Wang et al. proposed IPCGAN [9], an architecture that successfully generates new synthetic face images and preserves identities in specific age groups. It generates realistic, age-appropriate faces and guarantees that the synthesized faces have the same identity as the input image. In the experimental results, our system is compared with IPCGAN. Since IPCGAN only changes the facial lines, it cannot predict the appearance of children when they grow up.

In addition, HRFAE [8], proposed by Yao et al., combines age labels and latent vectors and can be used for face age editing on high-resolution images. The core idea is to create a latent space containing face identities and a feature modulation layer corresponding to the individual's age and then combine these two elements so that the generated output image is the specified target age. In the experimental results, our system is compared with HRFAE. Because HRFAE only considers facial lines, it cannot predict the appearance of children when they grow up.

3. Method

We propose a Face Prediction System for Missing Children, whose purpose is to predict children's future faces. It allows parents to quickly and inexpensively confirm the possibility of blood relation with any child. When parents search for missing children, our system helps to eliminate low similarity matches and narrow the search. Our system considers the respective features of the following two face images to predict the future face, including face images of the child before the disappearance and face images of the blood relatives. Our system combines StyleGAN2 and FaceNet methods to achieve prediction. StyleGAN2 is used to style mix two face images. FaceNet is used to compare the similarity of two face images. The input is an image of the missing child available before the disappearance and multiple images of family members related by blood. The output is a prediction result. More details can be found in Sections 3.1–3.4.

At the application level, we apply our Face Prediction System for Missing Children and the issues of searching for missing children to the concepts of IoE and AIoT, as shown in Figure 4, which will be described in detail below.

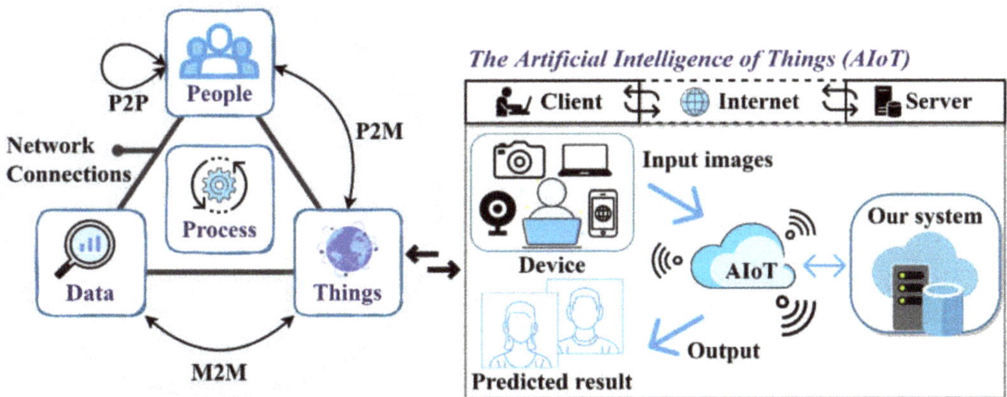

Figure 4. Our system is applied to the concepts of IoE and AIoT [53,54].

On the left side of Figure 4 is the IoE that combines machine-to-machine (M2M), people-to-people (P2P), and people-to-machine (P2M) connections. The difference between IoE and IoT is that IoT only focuses on the pillar of things, while IOE includes four pillars, namely things, people, process and data. IoE is the intelligent connection of the four pillars. The definition of the process is to provide the right information to the right person or machine at the right time to make the connection between people, things, and data more valuable. M2M is defined as the transmission of data from one machine or thing to another machine, including sensors, robots, computers, mobile devices, etc. These M2M connections can be considered IoT. P2P is defined as the transmission of information from one person to another. At present, P2P is mainly realized by mobile devices (such as PCs, TVs, and smartphones) and social networks (such as Facebook, Twitter, and LinkedIn). P2M is defined as the transmission of information between people and machines. People conduct complex data analysis through machines to obtain useful key information and help people make informed decisions. The following will explain the application mode of our system with the concepts of M2M, P2P and P2M.

- M2M: The user uses a device with a photographing function to transmit the photos of the children before going missing and the photos of their relatives to our system through the cloud network to predict the faces of the missing children and finally transmit the prediction results to the user through the cloud network (the above process can also be regarded as AIoT, as shown on the right side of Figure 4);
- P2P: Family members and friends of the missing children or police can publish relevant information about the missing children (including the time of disappearance, the place of disappearance, the photos before the disappearance and the predicted images from our system, etc.) to social media through mobile devices and social networks, hoping to be known and shared by netizens. The aim is to find the witnesses of the incident or people who know the context of the incident, who will provide relevant information to their families or police to assist in the arrest of the murderer;
- P2M: Family members, friends or police officers of missing children can use our system to predict the face of missing children at their present age and use the prediction results as one of the clues. Then, they can spread the image through TV, newspapers, magazines and various social media to let more people know about the case, and let people recall and judge whether they have seen this person. Finally, if people have clues, they can provide criminal clues to the police to help solve the case.

Our system mainly combines StyleGAN2 and FaceNet methods. StyleGAN2 is used to mix two images, and FaceNet is used to compare the similarity of the two images. The architecture of this system will be described in detail below.

3.1. Overview of the System Architecture

Figure 5 shows a flowchart of the image processing steps, divided into three main parts: data preprocessing, phase 1: filtering the best new face image, and phase 2: predicting the age and appearance of a missing child.

- Data preprocessing: This is used to take a single face image from the original image and output the dlatents for each face image. We will need the dlatents of each face image to use StyleGAN2 for face mixing. At the beginning of the first and second phases, we load the dlatents, and then proceed to mix the two images;
- Phase 1—filtering the best new face image: This is used to mix the two relatives with the highest similarity with the missing child. The face mixing result will have the appearance characteristics of the above two relatives. Finally, the system will select the best new face from multiple mixing results;
- Phase 2—predicting the age and appearance of a missing child: This is used to mix the best new face and the image of the missing child so that the face mixing result has not only the appearance characteristics of the missing child but also the appearance characteristics of the above two relatives. Finally, the system will select the best prediction result from multiple mixed results.

The details of these three parts will be described in order below.

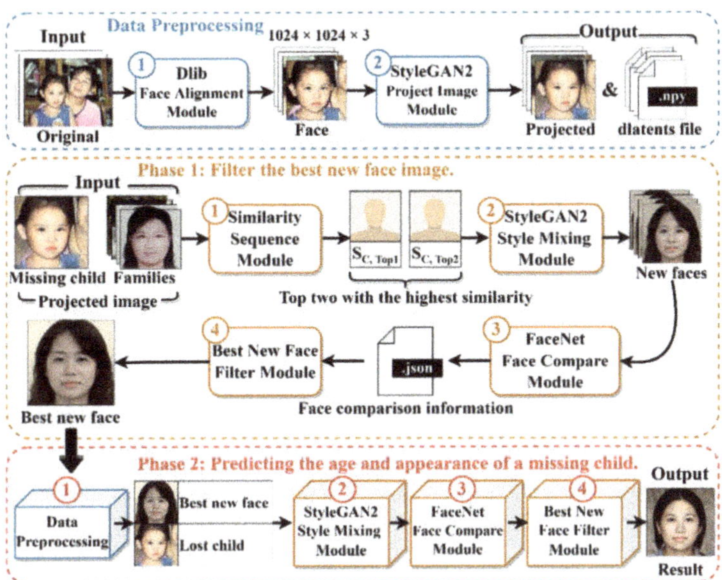

Figure 5. Flowchart of image processing steps.

3.2. Data Preprocessing

Data preprocessing consists of the following two steps:

1. Dlib Face Alignment Module: This module aligns and crops each face in the missing child's available image;
2. StyleGAN2 Project Image Module: Each face image is subjected to StyleGAN2 projection processing, and finally, the projected image and dlatents file are obtained. These data will be required as input during the first and second phases.

3.2.1. Dlib Face Alignment Module

Figure 6 shows the flowchart of the Dlib Face Alignment Module, which mainly corrects and truncates each face in the original image and outputs it as a face image of size 1024×1024. The Dlib Face Alignment Module contains three functions: 'Face Detector', 'Facial Landmark Predictor' and 'Face Alignment'.

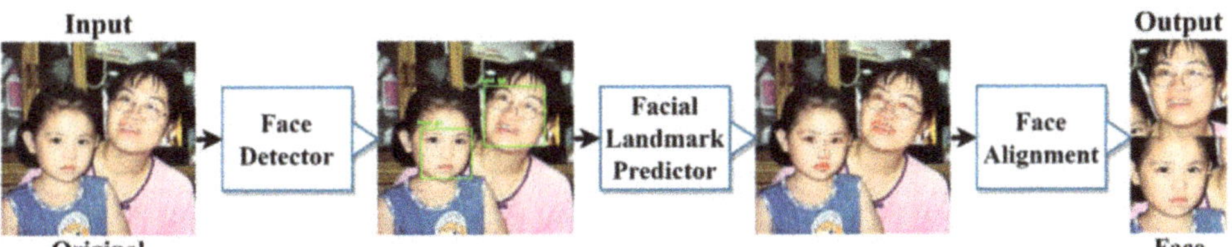

Figure 6. Dlib Face Alignment Module Flowchart.

- Face Detector: Each face in the original is detected and labelled with a number;
- Facial Landmark Predictor: The 68 landmarks of each face are predicted;
- Face Alignment: The image is rotated so that the landmarks of the eyes are horizontally aligned, then the face image is captured and the image is resized to 1024×1024.

3.2.2. StyleGAN2 Project Image Module

Figure 7 shows the StyleGAN2 projection process, whose input is an image of a missing child (the child in Figure 6). The *Projection image* is a function provided by StyleGAN2, which can iterate continuously on an input image (missing child face), producing a very similar one (a projection). In Figure 6, the projected image when iteration is 1 is the projected image after StyleGAN2 training, which is the default image of StyleGAN2. When iterating the 1000th time, the result of the projected image is very similar to the input image (missing child face), so we stored the dlatents this time as a NumPy file to be used for StyleGAN2 style mixing or interpolation in the future.

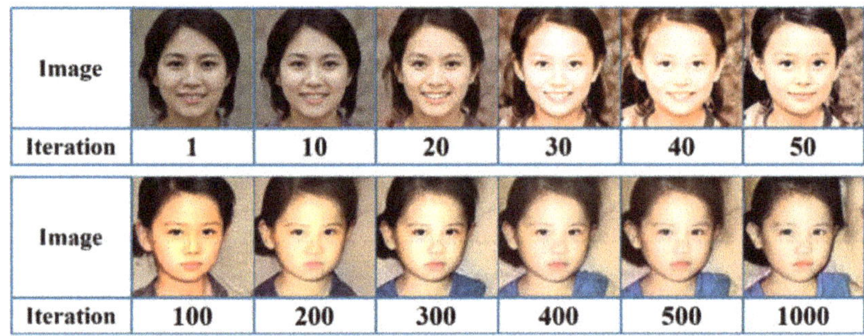

Figure 7. StyleGAN2 projection image process.

3.3. Phase 1: Filter the Best New Face Image

The first phase of the system is mainly to filter the best new face images. The input data is a projected image of a missing child and several projected images of family members. The output is an image of the best new face, one of the 36 mixed faces. There are four modules in the first processing phase: the Similarity Sequence Module, StyleGAN2 Style Mixing Module, FaceNet Face Compare Module and Best New Face Filter Module.

3.3.1. Similarity Sequence Module

The Similarity Sequence Module focuses on selecting the two family members most similar to the child from multiple family members. The missing child is first compared with each family member using FaceNet. All the similarities are ranked in descending order, and the images of the top two family members with the highest similarities are output.

3.3.2. StyleGAN2 Style Mixing Module

The StyleGAN2 Style Mixing Module inputs the top two dlatents with the highest similarity in the first phase, and after the StyleGAN2 style mixing process, a total of 36 mixed new faces will be generated. For example, Figure 8 is the StyleGAN2 style mixing result.

Figure 8. StyleGAN2 Style Mixing result.

3.3.3. Best New Face Filter Module

The Best New Face Filter Module mainly filters one of the 36 new faces mixed by StyleGAN2 as the best new face. In the first phase, this module mainly uses the similarity percentage metric to evaluate the advantages and disadvantages of 36 new faces. These 36 new faces $\langle n_1, n_2, n_3, \cdots, n_{36} \rangle$ will get a weight W_{1,n_k}, respectively. Then the system will rank W_{1,n_k} from small to large, and the minimum $\min W_{1,n_k}$ is the best new face.

$$W_{1,n_k} = W_{1,P_{n_k}} + W_{1,S_{n_k}} \tag{1}$$

$$W_{1,P_{n_k}} = \left| P_{C,Top1} - P_{n_k,Top1} \right| = \left| \frac{S_{C,Top1}}{S_{C,Top1} + S_{C,Top2}} - \frac{S_{n_k,Top1}}{S_{n_k,Top1} + S_{n_k,Top2}} \right| \tag{2}$$

$$W_{1,S_{n_k}} = \left| S_{C,Top1} - S_{n_k,Top1} \right| + \left| S_{C,Top2} - S_{n_k,Top2} \right| \tag{3}$$

Here, Equation (1) is the formula for weight W_{1,n_k} (refer to Table 2 for symbolic meaning), which mainly calculates the similarity percentage of $W_{1,P_{n_k}}$ and similarity of $W_{1,S_{n_k}}$ between children and 36 new faces. Then, $W_{1,P_{n_k}}$ and $W_{1,S_{n_k}}$ is added and called the Similarity Percentage Metric. Equation (2) is the similarity percentage $W_{1,P_{n_k}}$, which mainly calculates $P_{C,Top1}$ and $P_{C,Top1}$. The smaller the gap between the two, the better, indicating that the percentage value of the two is closer. Equation (3) is the similarity of $W_{1,S_{n_k}}$; the smaller the formula, the better, indicating that the new face image is more similar to the family. After calculating W_{1,n_k}, the system will sort each weight, and the smallest weight $\min W_{1,n_k}$ is the best new face.

Table 2. Symbol Definition.

Symbol	Meaning
n_k	$\langle n_k \rangle_{k=1}^{36} = \langle n_1, n_2, n_3, \cdots, n_{36} \rangle$, where n represents the generated new face image, and k is the item.
W_1 and W_2	W_1 represents the weight value of the first phase; W_2 represents the weight value of the second phase.
$S_{x,y}$	Represents the similarity comparison value of x and y. $0 \leq x, y \leq 100$.
$P_{x,y1}$ or $P_{x,y2}$	Represents the similarity proportion value of $S_{x,y1}$ or $S_{x,y2}$ among $S_{x,y1}$ and $S_{x,y2}$.
$Top1$ and $Top2$	$Top1$ Family members with the highest similarity to the child; $Top2$ Family members with the second-highest similarity to the child.
C	Missing child.
B	Best new face.

3.4. Phase 2: Predicting the Age and Appearance of a Missing Child

The second phase of the system focuses on predicting the current age of the missing child. The input data are the best new face and the missing child image, and the output data are the prediction result. A total of four modules were used in the second phase of processing, in the order of Data Preprocessing, StyleGAN2 Style Mixing Module, FaceNet Face Compare Module and Best New Face Filter Module. The two modules, Data Preprocessing and StyleGAN2 Style Mixing Module, operate in the same way as the corresponding modules in Phase 1, while the other modules are different.

3.4.1. FaceNet Face Compare Module

The FaceNet Face Compare Module mainly compares the best new face and the missing child image with each new face in the second phase. Finally, it records the similarity comparison information in the JSON file for subsequent analysis.

3.4.2. Best New Face Filter Module

The Best New Face Filter Module mainly selects the best prediction result from 36 new faces in the second phase. In the second phase, the similarity percentage is mainly used to evaluate the advantages and disadvantages of 36 new faces. These 36 new faces $\langle n_1, n_2, n_3, \cdots, n_{36} \rangle$ will get a weight W_{2,n_k}, respectively. Then the system will rank W_{2,n_k} from small to large, and the minimum $\min W_{2,n_k}$ is the best new face.

$$W_{2,n_k} = \left| S_{C,B} - S_{n_k,B} \right| \tag{4}$$

Equation (4) is the formula for weight W_{2,n_k} (refer to Table 2 for symbolic meaning), which is the similarity gap between 36 new faces and the best new face. The smaller W_{2,n_k} is, the more it means that the new face will be more similar to the best new face. After calculating W_{2,n_k}, the system will sort each weight and the smallest weight $\min W_{2,n_k}$ is the best new face.

4. Experiment

Figure 9 shows the experimental results of this system. The input data for this experiment were obtained from members of the same family. The first column in Figure 9 is the image of missing children. These three images are of different people; they are about 3 years old. The second column is images of family members or relatives of the missing children; the third column is our system, which contains the first and second phases. The input in the first phase is an image of the missing child available before their disappearance and multiple images of family members who are related by blood (dotted box in Figure 9), and the output is the best new face image, which is a mixed image of the facial features of two blood relatives. The input in the second phase is an image of the missing child available before their disappearance and an image of the best new face. The fourth column is the predicted results (output) of our system; the fifth column is the similarity comparison between the predicted result and the expected result, and the sixth column is the expected output, which is the ground truth, the faces of the missing children at the age of 20. The system mainly uses the face compare function provided by SKEye [55] for similarity comparison, and its similarity comparison refers to Algorithm 1. The predictions of the three sisters were compared to the expected output, and the results were 77%, 76% and 77%, respectively.

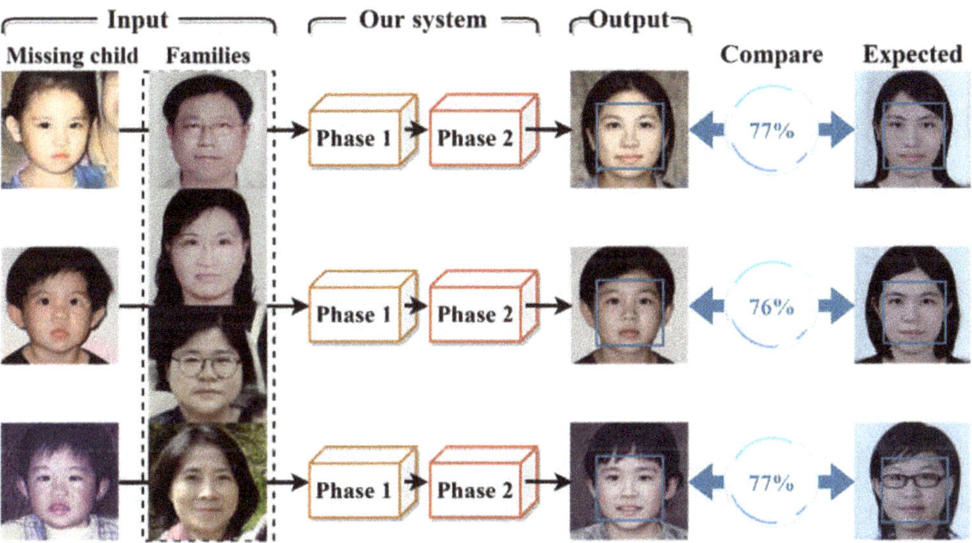

Figure 9. A comparison experiment of the similarity between the predicted image and the expected image.

From the physical appearance, it is difficult for humans to identify the gender of children under the age of three. We observe the child in the second row in Figure 9; she looks like a male, and the predicted image also looks like a male, but this does not affect the final similarity comparison of our prediction system. Because our system excludes human subjective judgments (including hairstyles) and only compares the similarity of facial features, the system will not be misled by physical appearance.

Algorithm 1 SKFace [55] Feature Comparison

Input: F_1: Features of the first face; F_2: Features of the second face;
Output: S: Similarity between F_1 and F_2;
1: Load F_1 and F_2;
2: Get F_1 and F_2 base64 code;
3: Verify whether F_1 and F_2 are recognized;
4: Calculate the distance between F_1 and F_2;
5: F_1 and F_2 are converted into similarity S.

Figure 10 shows the comparison diagram of our system, CAAE [6], HRFAE [8] and IPCGAN [9]. The first line is the input child image; Line 2 is the expected output; Lines 3~6 are the prediction results of the system, CAAE, HRFAE and IPCGAN and the similarity comparison results with the expected output. The similarity can correspond to Table 3. It can be seen from Figure 10 that, compared with other aging models, this system can produce more natural and high-resolution images, and the prediction accuracy is the highest, about more than 75%, which means that this system can well predict the appearance of children when they grow up.

Our system works for non-special families, direct blood relatives, and images with intact and undamaged faces. The following will list the conditions that do not apply because these reasons may result in low similarity:

- Special family, including half-brothers and half-sisters, etc.;
- Non-direct blood relatives, including an aunt's husband, uncle's wife and cousins, etc.;
- Incomplete or damaged face image, including poor image quality and face injuries, angles that are too skewed, expressions that are too exaggerated, etc.;
- Twins.

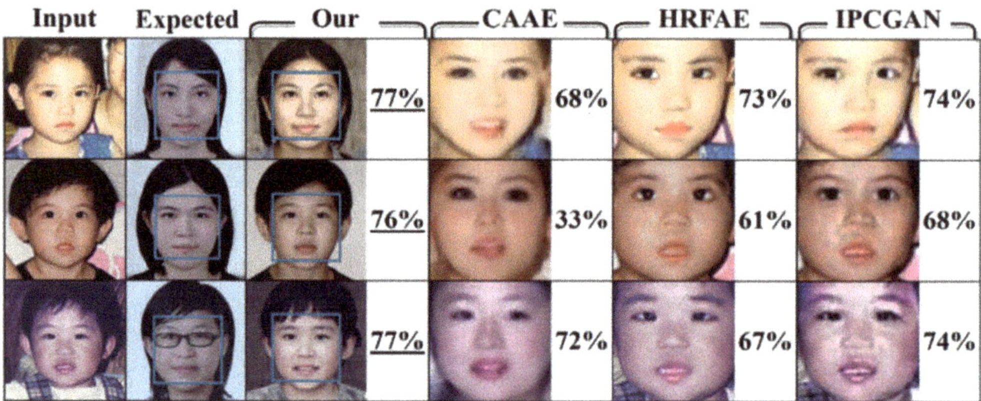

Figure 10. Comparison diagram of our system.

Table 3. Similarity comparison results of our system.

Input	Expected	Similarity between Input and Expected			
		Our	CAAE [6]	HRFAE [8]	IPCGAN [9]
		77%	68%	73%	74%
		76%	33%	61%	68%
		77%	72%	67%	74%

5. Conclusions

This study proposes a Face Prediction System for Missing Children, which can enable parents to quickly confirm whether they have the possibility of a parent-child relationship with any missing child, hoping to help parents find the missing child. The system combines FaceNet and StyleGAN2 methods to predict the appearance of missing children at their present age through similarity comparison and style mixing. Finally, we compare this system with other aging models, including CAAE, HRFAE and IPCGAN. Experiments show that this system has the highest prediction accuracy compared with other aging models, and the prediction results are of higher picture quality and natural.

Author Contributions: Conceptualization, G.-J.H.; methodology, Z.-J.T. and D.-C.W.; software, Z.-J.T.; validation, Z.-J.T.; investigation, Z.-J.T. and D.-C.W.; resources, D.-C.W.; writing—original draft preparation, Z.-J.T. and G.-J.H.; writing—review and editing, Z.-J.T. and G.-J.H.; supervision, C.-C.C.; project administration, G.-J.H. All authors have read and agreed to the published version of the manuscript.

Funding: This research received no external funding.

Informed Consent Statement: Informed consent was obtained from all participants involved in the study.

Acknowledgments: This work was supported in part by the Ministry of Science and Technology (MOST) of Taiwan under Grants MOST 110-2221-E-218-002 and in part by the "Allied Advanced Intelligent Biomedical Research Center, STUST" from Higher Education Sprout Project, Ministry of Education, Taiwan, and in part by the Ministry of Science and Technology (MOST) of Taiwan under Grant MOST 110-2221-E-218-007.

Conflicts of Interest: The authors declare no conflict of interest.

References

1. Federal Bureau of Investigation, 2020 NCIC Missing Person and Unidentified Person Statistics. Available online: https://www.fbi.gov (accessed on 28 January 2022).
2. National Crime Agency, UK Missing Persons Unit. Available online: http://www.missingpersons.police.uk (accessed on 28 January 2022).
3. Mohanty, S.P.; Choppali, U.; Kougianos, E. Everything you wanted to know about smart cities: The Internet of things is the backbone. *IEEE Consum. Electron. Mag.* **2016**, *5*, 60–70. [CrossRef]
4. Lacinák, M.; Ristvej, J. Smart City, Safety and Security. *Procedia Eng.* **2017**, *192*, 522–527. [CrossRef]
5. Ristvej, J.; Lacinák, M.; Ondrejka, R. On Smart City and Safe City Concepts. *Mob. Netw. Appl.* **2020**, *25*, 836–845. [CrossRef]
6. Zhang, Z.; Song, Y.; Qi, H. Age Progression/Regression by Conditional Adversarial Autoencoder. In Proceedings of the IEEE Conference on Computer Vision and Pattern Recognition (CVPR), Honolulu, HI, USA, 21–26 July 2017; pp. 4352–4360. [CrossRef]
7. Palsson, S.; Agustsson, E.; Timofte, R.; Gool, L.V. Generative Adversarial Style Transfer Networks for Face Aging. In Proceedings of the IEEE/CVF Conference on Computer Vision and Pattern Recognition Workshops (CVPRW), Salt Lake City, UT, USA, 18–22 June 2018; pp. 2165–21658. [CrossRef]
8. Yao, X.; Puy, G.; Newson, A.; Gousseau, Y.; Hellier, P. High Resolution Face Age Editing. In Proceedings of the International Conference on Pattern Recognition (ICPR), Milan, Italy, 10–15 January 2021; pp. 8624–8631. [CrossRef]
9. Tang, X.; Wang, Z.; Luo, W.; Gao, S. Face Aging with Identity-Preserved Conditional Generative Adversarial Networks. In Proceedings of the IEEE/CVF Conference on Computer Vision and Pattern Recognition (CVPR), Salt Lake City, UT, USA, 18–23 June 2018; pp. 7939–7947. [CrossRef]
10. Tanner, J.M.; Davies, P.S. Clinical longitudinal standards for height and height velocity for North American children. *J. Pediatr.* **1985**, *107*, 317–329. [CrossRef]
11. Weitzman, J. Epigenetics: Beyond face value. *Nature* **2011**, *477*, 534–535. [CrossRef]
12. Marioni, R.E.; Belsky, D.W.; Deary, I.J. Association of facial ageing with DNA methylation and epigenetic age predictions. *Clin. Epigenet.* **2018**, *10*, 140. [CrossRef] [PubMed]
13. Richmond, S.; Howe, L.J.; Lewis, S.; Stergiakouli, E.; Zhurov, A. Facial Genetics: A Brief Overview. *Front. Genet.* **2018**, *9*, 462. [CrossRef] [PubMed]
14. Miko, I. Gregor Mendel and the principles of inheritance. *Nat. Educ.* **2008**, *1*, 134.
15. Bowler, P.J. The Mendelian Revolution: The Emergence of Hereditarian Concepts in Modern Science and Society. *Baltim. J. Hist. Biol.* **1989**, *24*, 167–168.
16. The Human Life Cycle. Available online: https://med.libretexts.org/@go/page/1918 (accessed on 14 February 2022).
17. Goodfellow, I.J.; Pouget-Abadie, J.; Mirza, M.; Bing, X.; Bengio, Y. Generative adversarial networks. In Proceedings of the Neural Information Processing Systems (NeurIPS), Montreal, QC, Canada, 8–13 December 2014; pp. 2672–2680.
18. Karras, T.; Aila, T.; Laine, S.; Lehtinen, J. Progressive growing of gans for improved quality, stability, and variation. In Proceedings of the International Conference on Learning Representations (ICLR), Vancouver, BC, Canada, 30 April–3 May 2018.
19. Karras, T.; Laine, S.; Aila, T. A style-based generator architecture for generative adversarial networks. In Proceedings of the IEEE/CVF Conference on Computer Vision and Pattern Recognition (CVPR), Long Beach, CA, USA, 15–20 June 2019; pp. 4401–4410. [CrossRef]
20. Karras, T.; Laine, S.; Aittala, M.; Hellsten, J.; Lehtinen, J.; Aila, T. Analyzing and improving the image quality of stylegan. In Proceedings of the IEEE/CVF Conference on Computer Vision and Pattern Recognition (CVPR), Seattle, WA, USA, 13–19 June 2020; pp. 8107–8116. [CrossRef]
21. Abdal, R.; Qin, Y.; Wonka, P. Image2stylegan: How to embed images into the stylegan latent space? In Proceedings of the IEEE/CVF International Conference on Computer Vision, Seoul, Korea, 27–28 October 2019; pp. 4431–4440. [CrossRef]
22. Abdal, R.; Qin, Y.; Wonka, P. Image2stylegan++: How to edit the embedded images? In Proceedings of the IEEE/CVF Conference on Computer Vision and Pattern Recognition (CVPR), Seattle, WA, USA, 13–19 June 2020. [CrossRef]
23. Zhu, P.; Abdal, R.; Qin, Y.; Wonka, P. Sean: Image synthesis with semantic region-adaptive normalization. In Proceedings of the IEEE/CVF Conference on Computer Vision and Pattern Recognition, Seattle, WA, USA, 13–19 June 2020. [CrossRef]
24. Collins, E.; Bala, R.; Price, B.; Susstrunk, S. Editing in style: Uncovering the local semantics of gans. In Proceedings of the IEEE/CVF Conference on Computer Vision and Pattern Recognition (CVPR), Seattle, WA, USA, 13–19 June 2020.

25. Abdal, R.; Zhu, P.; Mitra, N.; Wonka, P. StyleFlow: Attribute-Conditioned Exploration of StyleGAN-Generated Images Using Conditional Continuous Normalizing Flows. *ACM Trans. Graph. (TOG)* **2021**, *40*, 1–21. [CrossRef]
26. Richardson, E.; Alaluf, Y.; Patashnik, O.; Nitzan, Y.; Azar, Y.; Shapiro, S.; Cohen-Or, D. Encoding in Style: A StyleGAN Encoder for Image-to-Image Translation. In Proceedings of the IEEE/CVF Conference on Computer Vision and Pattern Recognition (CVPR), Nashville, TN, USA, 20–25 June 2021; pp. 2287–2296. [CrossRef]
27. Patashnik, O.; Wu, Z.; Shechtman, E.; Cohen-Or, D.; Lischinski, D. StyleCLIP: Text-Driven Manipulation of StyleGAN Imagery. *arXiv* **2021**, arXiv:2103.17249.
28. Kim, H.; Choi, Y.; Kim, J.; Yoo, S.; Uh, Y. Exploiting Spatial Dimensions of Latent in GAN for Real-time Image Editing. In Proceedings of the IEEE/CVF Conference on Computer Vision and Pattern Recognition (CVPR), Nashville, TN, USA, 20–25 June 2021; pp. 852–861. [CrossRef]
29. Schroff, F.; Kalenichenko, D.; Philbin, J. FaceNet: A Unified Embedding for Face Recognition and Clustering. In Proceedings of the IEEE/CVF Conference on Computer Vision and Pattern Recognition (CVPR), Boston, MA, USA, 7–12 June 2015; pp. 815–823. [CrossRef]
30. Huang, G.B.; Ramesh, M.; Berg, T.; Learned-Miller, E. Labeled Faces in the Wild: A Database for Studying Face Recognition in Unconstrained Environments. Available online: http://vis-www.cs.umass.edu/lfw/lfw.pdf (accessed on 28 January 2022).
31. Baltrušaitis, T.; Robinson, P.; Morency, L.-P. OpenFace: An open source facial behavior analysis toolkit. In Proceedings of the IEEE Winter Conference on Application of Computer Vision, Lake Placid, NY, USA, 7–10 March 2016; pp. 1–10. [CrossRef]
32. Taigman, Y.; Yang, M.; Ranzato, M.; Wolf, L. DeepFace: Closing the Gap to Human-Level Performance in Face Verification. In Proceedings of the IEEE/CVF Conference on Computer Vision and Pattern Recognition (CVPR), Columbus, OH, USA, 23–28 June 2014; pp. 1701–1708. [CrossRef]
33. Parkhi, O.; Vedaldi, A.; Zisserman, A. Deep Face Recognition. In Proceedings of the British Machine Vision Conference (BMVC), Nottingham, UK, 1–5 September 2014; pp. 41.1–41.12. [CrossRef]
34. Cao, Q.; Shen, L.; Xie, W.; Parkhi, O.; Zisserman, A. VGGFace2: A dataset for recognising faces across pose and age. In Proceedings of the IEEE International Conference on Automatic Face & Gesture Recognition (FG 2018), Xi'an, China, 15–19 May 2018; pp. 67–74. [CrossRef]
35. Sun, Y.; Wang, X.; Tang, X. Deep Learning Face Representation from Predicting 10,000 Classes. In Proceedings of the IEEE Conference on Computer Vision and Pattern Recognition (CVPR), Columbus, OH, USA, 23–28 June 2014; pp. 1891–1898. [CrossRef]
36. Sun, Y.; Wang, X.; Tang, X. Deep Learning Face Representation by Joint Identification-Verification. In Proceedings of the 27th International Conference on Neural Information Processing Systems—Volume 2 (NIPS'14), Montreal, QC, Canada, 8–13 December 2014; pp. 1988–1996.
37. Sun, Y.; Wang, X.; Tang, X. Deeply learned face representations are sparse, selective, and robust. In Proceedings of the IEEE Conference on Computer Vision and Pattern Recognition (CVPR), Boston, MA, USA, 7–12 June 2015; pp. 2892–2900. [CrossRef]
38. Sun, Y.; Liang, D.; Wang, X.; Tang, X. DeepID3: Face Recognition with Very Deep Neural Networks. *arXiv* **2015**, arXiv:1502.00873.
39. Deng, J.; Guo, J.; Xue, N.; Zafeiriou, S. ArcFace: Additive Angular Margin Loss for Deep Face Recognition. In Proceedings of the IEEE/CVF Conference on Computer Vision and Pattern Recognition (CVPR), Long Beach, CA, USA, 15–20 June 2019. [CrossRef]
40. King, D.E. Dlib-ml: A Machine Learning Toolkit. *JMLR* **2009**, *10*, 1755–1758.
41. Yan, M.; Zhao, M.; Xu, Z.; Zhang, Q.; Wang, G.; Su, Z. VarGFaceNet: An Efficient Variable Group Convolutional Neural Network for Lightweight Face Recognition. In Proceedings of the IEEE/CVF International Conference on Computer Vision Workshop (ICCVW), Seoul, Korea, 27–28 October 2019; pp. 2647–2654. [CrossRef]
42. Kingma, D.P.; Welling, M. Auto-Encoding Variational Bayes. *arXiv* **2021**, arXiv:1312.6114.
43. Makhzani, A.; Shlens, J.; Jaitly, N.; Goodfellow, I.; Frey, B. Adversarial autoencoders. *arXiv* **2015**, arXiv:1511.05644.
44. Balasundaram, P.; Avulakunta, I.D. Human Growth and Development. In *StatPearls [Internet]*; StatPearls Publishing: Treasure Island, FL, USA, 2022. Available online: https://www.ncbi.nlm.nih.gov/books/NBK567767/ (accessed on 14 February 2022).
45. Bian, X.; Li, J. Conditional adversarial consistent identity autoencoder for cross-age face synthesis. *Multimed. Tools Appl.* **2021**, *80*, 14231–14253. [CrossRef]
46. Zhu, J.; Park, T.; Isola, P.; Efros, A.A. Unpaired Image-to-Image Translation using Cycle-Consistent Adversarial Networks. In Proceedings of the IEEE International Conference on Computer Vision, Venice, Italy, 22–29 October 2017; pp. 2242–2251. [CrossRef]
47. Isola, P.; Zhu, J.; Zhou, T.; Efros, A.A. Image-to-Image Translation with Conditional Adversarial Networks. In Proceedings of the IEEE Conference on Computer Vision and Pattern Recognition (CVPR), Honolulu, HI, USA, 21–26 July 2017; pp. 5967–5976. [CrossRef]
48. Choi, Y.; Choi, M.; Kim, M.; Ha, J.-W.; Kim, S.; Choo, J. StarGAN: Unified Generative Adversarial Networks for Multi-Domain Image-to-Image Translation. In Proceedings of the IEEE/CVF Conference on Computer Vision and Pattern Recognition (CVPR), Salt Lake City, UT, USA, 18–23 June 2018; pp. 8789–8797. [CrossRef]
49. Shen, Y.; Gu, J.; Tang, X.; Zhou, B. Interpreting the Latent Space of GANs for Semantic Face Editing. In Proceedings of the IEEE/CVF Conference on Computer Vision and Pattern Recognition (CVPR), Seattle, WA, USA, 13–19 June 2020; pp. 9240–9249. [CrossRef]
50. Mirza, M.; Osindero, S. Conditional generative adversarial nets. *arXiv* **2014**, arXiv:1411.1784.

51. Roberts, A.M. *The Complete Human Body: The Definitive Visual Guide*; DK: London, UK, 2016.
52. Polan, E.; Taylor, D. *Journey across the Life Span: Human Development and Health Promotion*; F.A. Davis Co.: Philadelphia, PA, USA, 1998.
53. The Internet of Everything How More Relevant and Valuable Connections Will Change the World. Available online: https://www.cisco.com/c/dam/global/en_my/assets/ciscoinnovate/pdfs/IoE.pdf (accessed on 28 January 2022).
54. Session 2: Pillars of the IoE. Available online: https://www.open.edu/openlearn/mod/oucontent/view.php?id=48819 (accessed on 28 January 2022).
55. SKEye, Face Compare. Available online: https://www.sk-ai.com/Experience/face-compare (accessed on 28 January 2022).

 *electronics*

Article

Feature Extraction of Anomaly Electricity Usage Behavior in Residence Using Autoencoder

Chia-Wei Tsai [1], Kuei-Chun Chiang [1,2], Hsin-Yuan Hsieh [1], Chun-Wei Yang [3,4], Jason Lin [5] and Yao-Chung Chang [1,*]

[1] Department of Computer Science and Information Engineering, National Taitung University, No. 369, Section 2, University Road, Taitung 95092, Taiwan; cwtsai@nttu.edu.tw (C.-W.T.); chun@iii.org.tw (K.-C.C.); 0911137@gm.nttu.edu.tw (H.-Y.H.)
[2] Digital Transformation, Institute for Information Industry, 11F, No. 106, Section 2, Heping E. Road, Taipei 106, Taiwan
[3] Center for General Education, China Medical University, No. 100, Section 1, Jingmao Road, Beitun District, Taichung 406040, Taiwan; cwyang@mail.cmu.edu.tw
[4] Master Program for Digital Health Innovation, College of Humanities and Sciences, China Medical University, No. 100, Section 1, Jingmao Road, Beitun District, Taichung 406040, Taiwan
[5] Department of Computer Science and Engineering, National Chung Hsing University, No. 145, Xingda Road, South District, Taichung 40227, Taiwan; jasonlin@nchu.edu.tw
* Correspondence: ycc@nttu.edu.tw

Abstract: Due to the climate crisis, energy-saving issues and carbon reduction have become the top priority for all countries. Owing to the increasing popularity of advanced metering infrastructure and smart meters, the cost of acquiring data on residential electricity consumption has substantially dropped. This change promotes the analysis of residential electricity consumption, which features both small and complicated consumption behaviors, using machine learning to become an important research topic among various energy saving and carbon reduction measures. The main subtopic of this subject is the identification of abnormal electricity consumption behaviors. At present, anomaly detection is typically realized using models based on low-level features directly collected by sensors and electricity meters. However, due to the significant number of dimensions and a large amount of redundant information in these low-level features, the training efficiency of the model is often low. To overcome this, this study adopts an autoencoder, which is a deep learning technology, to extract the high-level electricity consumption information of residential users to improve the anomaly detection performance of the model. Subsequently, this study trains one-class SVM models for anomaly detection by using the high-level features of five actual residential users to verify the benefits of high-level features.

Keywords: energy saving; carbon reduction; advanced metering infrastructure; low-voltage users; anomaly detection; autoencoder

Citation: Tsai, C.-W.; Chiang, K.-C.; Hsieh, H.-Y.; Yang, C.-W.; Lin, J.; Chang, Y.-C. Feature Extraction of Anomaly Electricity Usage Behavior in Residence Using Autoencoder. *Electronics* **2022**, *11*, 1450. https://doi.org/10.3390/electronics11091450

Academic Editor: Floriano De Rango

Received: 2 April 2022
Accepted: 28 April 2022
Published: 30 April 2022

Publisher's Note: MDPI stays neutral with regard to jurisdictional claims in published maps and institutional affiliations.

1. Introduction

Owing to the depletion of fossil energy and increasingly serious global warming problems, the effective decrease in fossil energy consumption and energy and carbon reduction has become a common concern for governments and enterprises around the world. According to the data from the Bureau of Energy, Ministry of Economic Affairs of Taiwan, sectors that consumed the most energy in Taiwan were the industrial (55.9%), service (17.7%), and residential sectors (17.6%). The data clearly indicated that the electricity consumption of the residential sector was the third highest, only slightly lower than that of the service sector. Therefore, if the electricity consumption of the residential sector can be effectively reduced, considerable energy-saving benefits will be achieved. However, unlike the industrial and service sectors that comprise medium and large users, the residential sector comprises a large number of small users (approximately 13.2 million non-business users and 1.03 million business users of low-voltage meters). In addition, the electricity consumption behaviors of different users vary significantly, complicating the development of a universal energy-saving strategy for residential users. Fortunately, in recent years, information and communication technology has developed rapidly; mobile devices, mobile networks, and Internet of Things (IoT) devices have gained a significant amount of popularity. Furthermore, power companies have actively promoted the advanced metering infrastructure (AMI) and smart meters to replace the existing mechanical meters for understanding the power-load behavior of low-voltage users quickly and efficiently.

The cost of collecting the electricity consumption data of low-voltage users decreases every year, and related energy management systems and devices are gradually implemented. Despite this, residential users lack the motivation to apply energy saving and carbon reduction measures and introduce home energy management systems (HEMS) due to the low electric valance in Taiwan. Therefore, some research [1–13] has begun using machine learning techniques to collect and analyze the electricity consumption data of residential users and establish artificial intelligence (AI) models to provide appropriate and tailored energy-saving suggestions. Among them, if a mechanism can identify the abnormal electricity consumption behavior of residential users and propose appropriate energy management or saving suggestions, it will be particularly effective in improving user motivation in terms of energy-saving measures. Therefore, various anomaly detection techniques for the energy consumption of residential users and buildings have been proposed and discussed.

At present, the studies predominantly use the low-level feature data (i.e., the row data without being extracted), including electricity consumption data and associated features (temperature and humidity, summer/non-summer months, working/non-working days, etc.) to train machine learning models. However, the significant number of dimensions and a large amount of redundant information of these low-level features can compromise the training performance of subsequent anomaly detection algorithms. Although techniques such as principal component analysis (PCA) and feature selection can be adopted to improve this issue, how to provide a more efficient solution is still an important research topic. Therefore, this study further discusses this topic. In summary, this study wants to improve a method to extract the essences of the low-level features by using a deep neural network-autoencoder. That is, this study wants to use the autoencoder to extract the high-level features (i.e., code in the autoencoder) from the low-level features of the electricity consumption data of residential users. As the high-level features can decode to the original low-level features and the dimensions of the high-level features are less than those of the low-level features, the high-level features are more representative of the power consumption behaviors of users. That is, the high-level features are the essence of the user's power consumption data. This study uses the actual power consumption data of the five resident users to execute the experiments to verify whether using the high-level features to train the anomaly detection algorithm can benefit performance over using the low-level features. We use an anomaly detection algorithm, one-class SVM, to train the two anomaly detection models with the high-level and low-level features, respectively, and then analyze the performances of the two models to verify the feasibility of the proposed high-level feature extraction method.

The remainder of this paper is organized as follows. Section 2 reviews the related literature and technologies, Section 3 describes the research methods and processes, and Section 4 reports the implementation of the function and result comparison. Section 5 provides a conclusion and recommendations for future research topics and directions.

2. Background

This chapter first reviews the relevant literature on anomaly detection and then briefly explains the machine learning techniques used in this study, particularly the autoencoder and one-class SVM.

2.1. Anomaly Detection

As early as the 19th century, the statistical community had already started detecting anomalies in data. Anomalies are also referred to as outliers, biases, inconsistencies, and exceptions [14]. Anomalies typically include (1) point, (2) contextual, and (3) collective anomalies. The detection of point anomalies is the most simple and common anomaly detection method and strategy. However, rather than a pattern, point anomalies often represent a noise and consequently possess a low practical value. Alternatively, contextual anomalies are typically analyzed in a specific time sequence and spatial data to determine abnormal behaviors in the specific context, whereas collective anomalies often analyze group data comprising multiple pieces and evaluate whether the resultant model is anomalous. The occurrence of contextual anomalies depends on the availability of contextual attributes in the data. Therefore, when point anomaly detection is supplemented with contextual anomaly detection or part of the group data are categorized as contextual attributes, both point and collective anomalies are considered equivalent to contextual anomalies. Consequently, during anomaly detection, most studies convert anomaly events to contextual anomalies for analysis and processing. Anomaly detection strategies can be divided, according to the inclusion of labels in the analysis datasets, into three types: (1) supervised, (2) unsupervised, and (3) semi-supervised.

The primary techniques adopted by the existing literature to detect abnormal power consumption behaviors include [15] (1) anomaly detection models based on regression models, (2) anomaly detection models based on classifiers, and (3) others. This section divides anomaly detection techniques for electricity consumption according to their type and provides a brief review and description.

Anomaly detection models based on regression models first train the regression model using historical power-related data and then use the model to predict future consumption. An anomaly is detected upon a large deviation between the predicted and actual values (for example, the actual value is greater than the predicted threshold). Zhang et al. [16] developed an abnormal electricity load detection model based on a linear regression model and used its predications as the baseline. Power consumption data were considered abnormal when either significantly lower or higher than the threshold. Although the study provided a load anomaly detection solution that incorporated environmental factors, it could not accurately identify anomalies for residential users owing to their sensitivity to temperature. In addition, as the model was only trained with environmental factors, it might be inapplicable in an environment with a constant annual temperature. Alternatively, Zhou et al. [17] proposed an anomaly detection model based on a hybrid prediction model. The hybrid model integrated the ARIMA model with the ANN model, compensating the prediction error of the former in nonlinear regression and providing the advantages of both linear and nonlinear models. Although this approach improved the prediction accuracy, the anomaly detection strategy used was excessively simple and required further improvement. To eliminate detection errors caused by simple detection methods, Luo et al. [18] proposed an anomaly detection model based on dynamic regression. Instead of a fixed threshold, the model could calculate a dynamic, adaptive threshold for the difference between the predicted and actual loads during anomaly detection. The proposed dynamic-detection rule could improve the accuracy of anomaly detection. However, because the study used the results of the prediction model as the only reference for anomaly detection, an independent detection mechanism was lacking for anomaly detection, risking a decrease in anomaly detection accuracy when the prediction value was inaccurate. Fenza et al. developed a drift-aware methodology for detecting anomalies in smart grids [19]. Historical data were used to train the long short-term memory (LSTM) and then to determine the anomaly detection thresholds from the prediction error trends obtained by the LSTM over time. As the study aimed to explore the abnormal load profile of users, the basis of anomaly detection was the error trend rather than the error between the predication and actual result for a specific time. Inayah et al. [20] used SARIMA and ANN models to predicate power consumption of the college buildings, and they adopted the difference between the actual and prediction values to identify the anomaly events. Then, the results of the experiment proved that the ANN model has a better performance than the SARIMA model. Additionally, it is noteworthy that this kind of anomaly detection technology can also be used to protect the cybersecurity issue. For example, Zhang et al. [21] proposed a robustness assessment framework for wind power, and they evaluated the performances of the six forecasting models in terms of protection against the false data injection attack.

Anomaly detection models based on classifiers can be further divided into supervised and unsupervised/semi-supervised models according to the type of classifier. Jokar et al. developed an anomaly detection model for power theft based on supervised learning [22]. During the training process, the k-means cluster analysis algorithm and silhouette coefficient determined the number of patterns in the dataset, and an SVM-based classifier learned the normal and abnormal patterns. Pinceti et al. [23] conducted a model comparison study, during which different supervised learning models detected abnormal load redistribution events. After comparing kNN, SVM, and RNN models, the study suggested that the performance of the kNN model was superior. Fang et al. [24] adopted the extreme learning machines and the ensemble learning strategy to design a supervised learning anomaly detection system for various users (i.e., the low-voltage non-resident, the low-voltage resident, the high-voltage resident, and the photovoltaic user). Wang et al. [25] proposed a semi-supervised learning anomaly detection model, sample efficient home power anomaly detection (SEPAD), in which the k-means and z-score function [26] were used to point out the suspicious data, and a semi-SVM based pattern matching algorithm was proposed to identify anomaly power consumption events. Hosseini et al. [27] focused on the appliance-level anomaly detection and trained the classification modes by using the operation patterns for the refrigerators depending on the semi-supervised learning strategy. Fan et al. [28] proposed a building electricity anomaly detection model based on unsupervised classification to reduce the training cost lower than that of supervised learning-based models. The study first determined the primary load frequency of users using spectral density analysis and features affecting the electricity consumption behavior using a decision tree, and then calculated the anomaly score of each event using the autoencoder, which is an unsupervised learning model, and ensemble learning. An event was defined as an anomaly if its anomaly score was higher than the preset threshold. Pereora et al. [29] developed an autoencoder-based unsupervised anomaly detection model for detecting anomalies in solar power generation. They also applied a variational self-attention mechanism to improve the performance of the autoencoder. Although anomaly detection techniques based on unsupervised learning do not require additional training to identify abnormal data, and therefore have a low training cost, evaluating their detection results is difficult due to the lack of reference labels [30]. Additional analysis (such as normal distribution analysis, data visualization analysis, and consulting domain experts) is often required to verify that the specific event is an anomaly.

Others include Janetzko et al. [31], who used the visual analysis to identify the anomaly power consumption events. The study [32] adopted the Hilbert-Huang transform and instantaneous frequency analysis to analyze the hidden anomaly events in commercial buildings. Cabrera et al. [33] adopted an anomaly detection method based on rule-based learning to analyze the waste of electricity in school buildings. They reduced the number of features using data mining methods and introduced various rules to identify wasteful behaviors. Li [34] uses statistical methods and clustering algorithms to identify the anomaly power consumption events in the short-term and long-term time scale data, respectively.

2.2. Autoencoder

An autoencoder [35,36] is an unsupervised learning algorithm in deep learning. The model is trained by defining the data (X) and output data (Y). According to the neural network architecture, an autoencoder comprises an encoder and decoder, which have neural networks with the same number of neurons. The encoder converts the input data into high-level features (Z) through the hidden layer, and the decoder reconstructs these high-level features into input data through the hidden layer. The autoencoder aims to restore the high-level features of the input data as much as possible using the decoder. Its loss function often uses mean squared error (MSE) or cross-entropy losses. Two common autoencoder structures exist: undercomplete autoencoders whose number of neurons in the hidden layer is smaller than or equal to that in the decoder, and overcomplete autoencoders whose number of neurons in the hidden layer is larger than or equal to that in the decoder. Basic autoencoder structures comprise three fully connected layers: an input, hidden, and output layer. Both the number of hidden layers and its number of neurons can be adjusted to improve the model performance.

2.3. One-Class SVM

One-class SVM is an unsupervised algorithm [37–39]. As the name suggests, it classifies incoming training data into one category. A decision boundary is first learned using the characteristics of these normal samples, which are then used to determine the similarity between the new and training data. Abnormal data are identified when they exceed the boundary. If the kernel function adopts the Gaussian Redial Basis (RBF), features of the training data are first projected to high dimensions and then projected back to the original data dimension once the largest segmentation platform, the hyperplane, is determined in the high dimension. The one-class SVM algorithm is similar to that of two-class SVM. The only difference is that the former searches for the hyperplane that contains all the normal training data instead of the hyperplane that splits training samples into two categories.

3. Research Methods

This chapter describes the research methods and processes (Figure 1). Data from a full year of electricity consumption of lower-voltage residential users were analyzed to assess whether the proposed method could effectively detect abnormal electricity consumption behaviors. The details of the research methods and process are listed below.

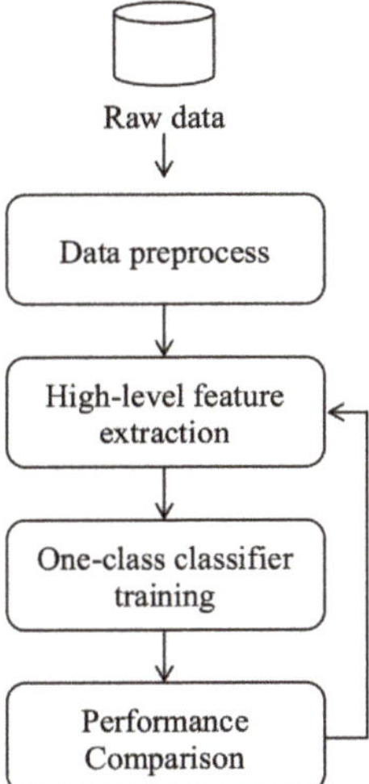

Figure 1. Research structure and process.

3.1. Data Preprocessing

Although the sources of the electricity consumption data of low-voltage users were predominantly smart meters and home energy management systems, the collected data could still have noises, outliers, or missing values owing to the noise and short-term failure of sensors during communication and data transmission. These noisy and abnormal data could affect the subsequent model training and performance and must therefore be filtered and processed prior to model training and analysis. As assigning missing value and restoring abnormal data (e.g., using the regression model) could compromise the accuracy of model establishment, abnormal and missing data were directly deleted in this study, that is, the electricity load data of a day was deleted upon the presence of any missing or abnormal value. Furthermore, because the international electric power industry routinely used 15 min as the sampling frequency of user electricity load data, this study resampled the original load data to 1 record/15 min using data averaging. During data preprocessing and after the removal of missing and abnormal values, to accelerate model training and increase model accuracy, a min–max normalization was performed to convert the data to a range of [0:1], the formula of which is:

$$s^i_{norm} = \frac{s^i_{original} - s_{min}}{s_{max} - s_{min}}$$

where $s^i_{original}$ is the i-th original sample data, s_{min} and s_{max} the minimum and maximum values of the original sample, respectively, and s^i_{norm} the normalized value of the i-th sample. Finally, the normalized data were then divided into a training and test dataset at a ratio of 80:20.

3.2. High-Level Feature Extraction

To facilitate the identification of abnormal electricity consumption behaviors, the study used the autoencoder for feature extraction. The autoencoder then encoded and compress the features of the 96 daily electricity load records to obtain low-dimensional electricity load features. This study adopted the multilayer undercomplete autoencoder as the primary high-level feature extraction model, which compressed and extracted the original low-level features (i.e., 96-dimensional features) into two-dimensional high-level features. Figures 2 and 3 present the model architecture, in which the intention of the first and the second dimensions of input and out shapes are the batch size and input size, and the term "None" in the first dimension means the batch size depends on how many samples we give for training.

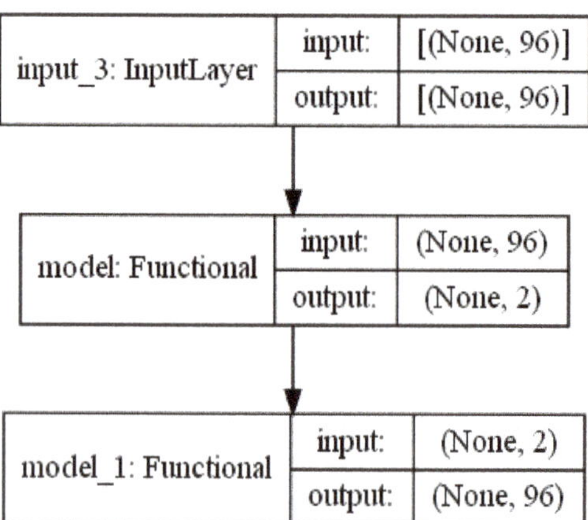

Figure 2. Structure of the autoencoder model.

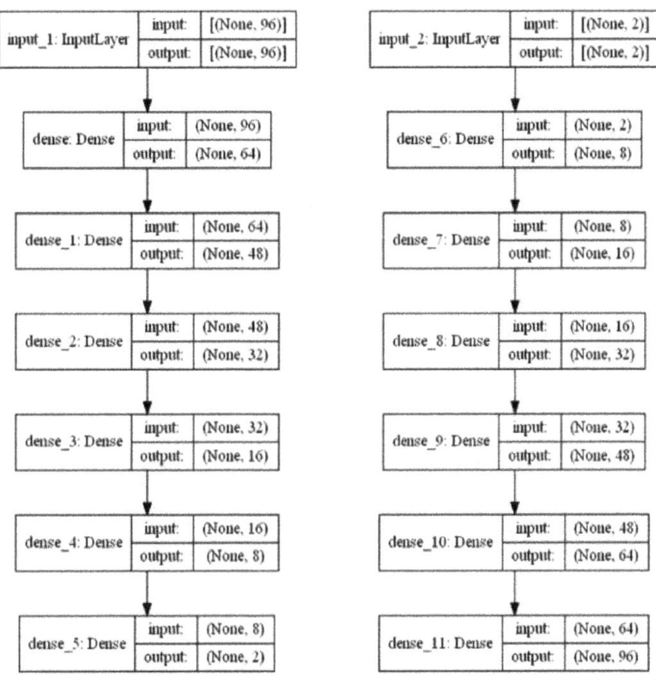

Figure 3. Structures of the encoder (left) and decoder (right).

Here, this study analyzes the time complexity of this network model to briefly obtain a time complexity formula. We know that the time complexity of matrix multiplication for $M_{ij} \times M_{jk}$ is $O(i \times j \times k)$. In the forward propagation, from the i-th layer to the $(i + 1)$-th layer, a matrix multiplication and an activation function must be computed. The time complexity of matrix multiplication is $O\left(n^i_{neuron} \times n^{i+1}_{neuron} \times n_{sample}\right)$, where n^i_{neuron} (n^{i+1}_{neuron}) denotes the number of neurons in the i-th ((i+1)-th) layer, and n_{sample} is the number of samples used to train the network. Due to the element-wise operation, the time complexity of the activation is $O\left(n^{i+1}_{neuron} \times n_{sample}\right)$. Therefore, the total time complexity is $O\left(n^i_{neuron} \times n^{i+1}_{neuron} \times n_{sample} + n^{i+1}_{neuron} \times n_{sample}\right) \approx O\left(n^i_{neuron} \times n^{i+1}_{neuron} \times n_{sample}\right)$. For all networks, the time complexity is $O\left(\sum_{i=1}^{n_{layer}} \left(n^i_{neuron} \times n^{i+1}_{neuron}\right) \times n_{sample}\right)$, where n_{layer} denotes the number of layers in the network model. In the backward propagation, from $(i + 1)$-th layer to the i-th layer, we must compute the error signal matrix by an element-wise multiplication operation, use a matrix multiplication to compute the delta weights, and then adjust the weights by using element-wise operation. Therefore, the total time complexity is $O\left(2 \times n^{i+1}_{neuron} \times n_{sample} + n^i_{neuron} \times n^{i+1}_{neuron} \times n_{sample} + n^i_{neuron} \times n^{i+1}_{neuron}\right) \approx O\left(n^i_{neuron} \times n^{i+1}_{neuron} \times n_{sample}\right)$ which is the same as the time complexity in the forward propagation. Thus, the total time complexity of the network model in both propagations is $O\left(\sum_{i=1}^{n_{layer}} \left(n^i_{neuron} \times n^{i+1}_{neuron}\right) \times n_{sample} \times n_{epoch}\right)$, where n_{epoch} denotes the number of training iteration.

3.3. One-Class Classifier Training

Once high-level feature extraction is completed, a one-class classifier trained the model to detect abnormal electricity consumption behaviors. As a type of classifier, the one-class classifier primarily uses single-class samples for model training, allowing the model to identify a new event and determine whether it belongs to the specific class of events. A positive result indicates that the new event belongs to the class whereas a negative result indicates that the new event does not belong to the class. The one-class classifier cannot provide further information on which class it belongs to. This study adopted the one-class SVM algorithm as the one-class classifier. As the proportion of the abnormal power consumption behaviors of general users was normally low, the study assumed that abnormal behaviors accounted for 2% of the overall load and used this assumption to train the model. In addition, to verify whether the proposed high-level feature extraction method could effectively escalate the anomaly detection efficiency of the one-class classifier, the study also used low-level features to train the anomaly detection model and compared its performance with that of the model trained using high-level features.

3.4. Performance Comparison

Owing to the lack of labels, anomalies detected by unsupervised strategies often had a low explanatory power. As insufficient evidence was available for proving whether the identified anomalies were true anomalies, domain experts should assist in the detection. Therefore, this study used data visualization to analyze the performances of models trained using high-level and low-level features. During data visualization, cluster analysis was adopted to determine the main load pattern of users among the electricity loads of normal consumption behaviors identified by the model. In addition, abnormal electricity consumption behaviors and characteristics identified by the two models were compared and their differences were analyzed to assess the pros and cons of the model trained using high-level features. The k-means++ algorithm was used during cluster analysis to determine the main electricity consumption characteristics of users, and a silhouette coefficient determined the appropriate number of clusters. Finally, the center point of each cluster was defined as the load characteristics of users, plotting its range using the Q1 and Q4 of the quartile to evaluate the usefulness of the abnormal electricity consumption detection model.

According to the steps and procedures, the study collected a full year of electricity consumption data of five congregate residences and performed data preprocess, high-level feature extraction, abnormal electricity consumption detection model training, main electricity consumption feature extraction, and performance analysis and evaluation, the results of which are described in the next chapter.

4. Results and Discussion

The data sources of this empirical research were five residences randomly selected from 200 residential users who had installed energy management systems. Data were collected at a frequency of one electricity consumption record per minute (for a total of 1440 records per day) between 1 January 2020 and 31 December 2020. The study then resampled the data to one record per 15 min (for a total of 96 records per day) using data averaging to match the main measurement unit adopted by power companies in Taiwan. Personal information was removed prior to data acquisition.

Subsequently, during model training, the epochs of the autoencoder was set to 10,000. MSE was chosen as the loss function owing to its sensitivity toward extreme values, and adaptive moment estimation (ADAS) was selected as the optimizer. Figure 4 presents the evaluation results of the trained model. According to the time complexity formula in Section 3.2, we can obtain the time complexity of each user's autoencoder model is approximately $O(2^{36})$.

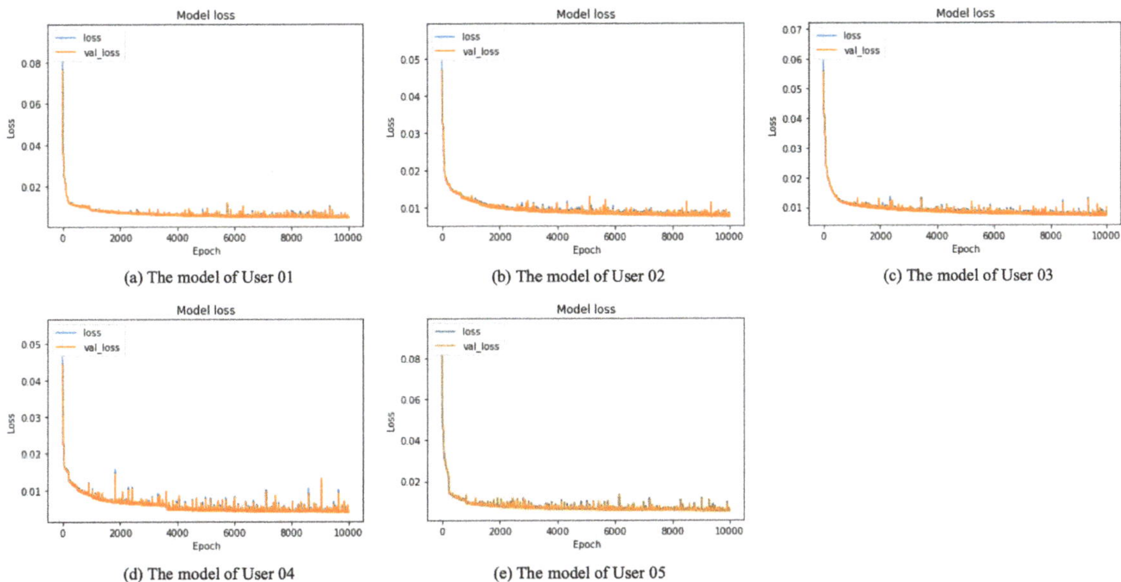

Figure 4. Learning curves of model loss in the five autoencoder models.

The anomaly detection model was then trained by the one-class SVM using the high-level and low-level features of five residences. The dates of all detected anomalies were labelled for subsequent visualization analysis. The study used the one-class SVM algorithm in the scikit-learn, version 1.0.2, during implementation. The kernel parameter of the model was set to linear, the gamma to auto, and all nu values to 0.02 (that is, anomalies accounted for 2% of the sample dataset). The remaining parameters were the default value. Next, the k-means of the cluster analysis, silhouette coefficient, and quartiles were calculated to plot the primary electricity load behavior of individual users. This was concluded by plotting the load profiles of the abnormal electricity consumption behaviors of users detected by high-level and low-level features as well as their main load behaviors to compare the performances of the two models.

Figures 5–9 show the anomaly power consumption events for the five tested residential users detected using high-level and low-level features. In the graphs, the green curve denotes the central value of the primary load behavior of users, the light green and grass green areas denote the Q1–Q4 range of the corresponding electricity consumption feature, and the red curve denotes the abnormal electricity consumption load. The plots on the left are anomalies detected using low-level features, whereas those on the right are anomalies detected using high-level features. The figures of the individual anomaly event are given in Appendix A.

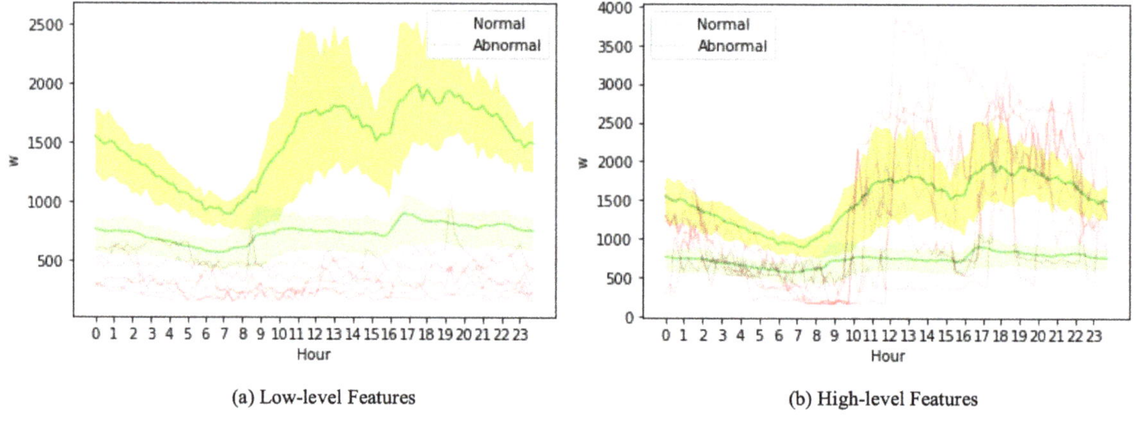

(a) Low-level Features   (b) High-level Features

Figure 5. Load profiles of the main and abnormal electricity consumption events of User 01.

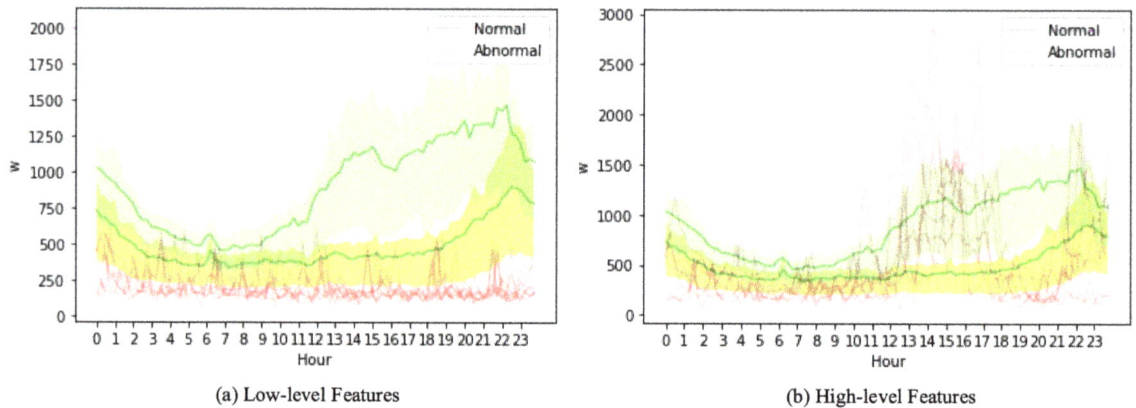

(a) Low-level Features   (b) High-level Features

Figure 6. Load profiles of the main and abnormal electricity consumption events of User 02.

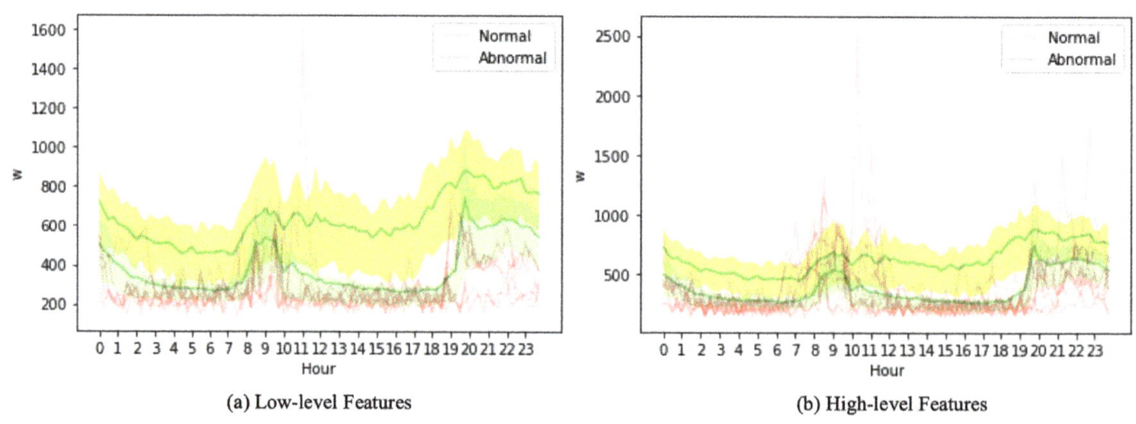

(a) Low-level Features   (b) High-level Features

Figure 7. Load profiles of the main and abnormal electricity consumption events of User 03.

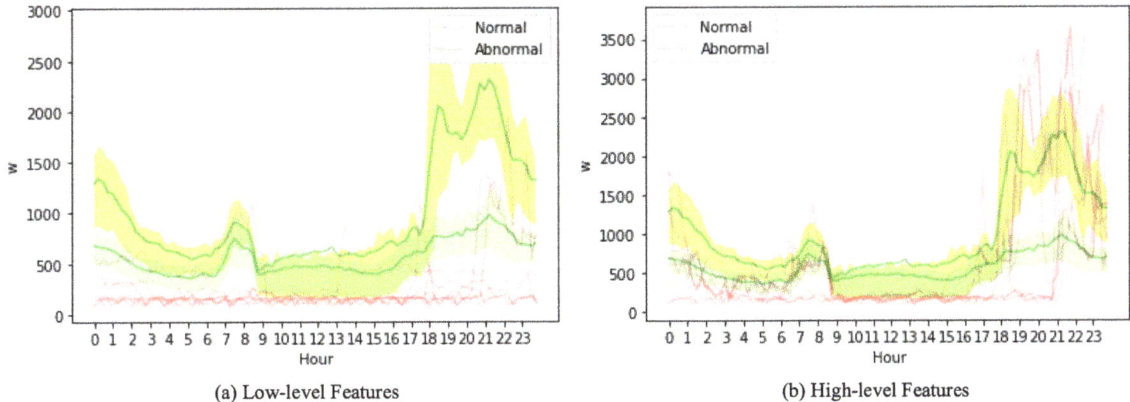

(a) Low-level Features

(b) High-level Features

Figure 8. Load profiles of the main and abnormal electricity consumption events of User 04.

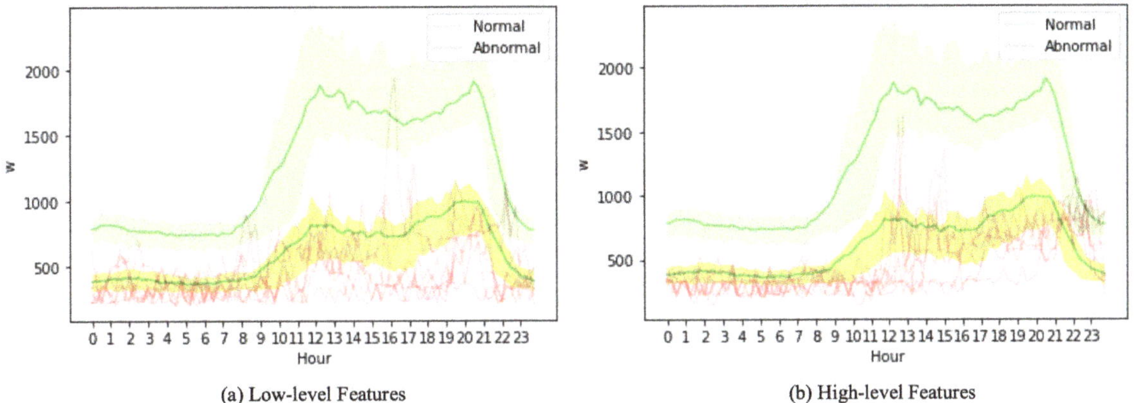

(a) Low-level Features

(b) High-level Features

Figure 9. Load profiles of the main and abnormal electricity consumption events of User 05.

Note that the electricity consumption behaviors of the five residents can be divided into two groups. An analysis of the primary difference between the two groups indicated that the major contributing factor was temperature, because high-load anomalies mostly occurred in summer months, and low-load anomalies in non-summer months. Here, the definition of summer and non-summer months is the same as that used by power companies in Taiwan, that is, summer months span from the start of June to the end of September, and the remaining months are non-summer months.

Visualization analysis clearly indicated that anomalies detected using high-level features were often extreme power consumption behaviors displayed as sharp rises or falls. These anomalies demonstrated significantly more distinctive features than those of the main load behaviors of the user, making it highly likely that they were real anomalies. In contrast, anomalies detected using low-level features were predominantly load behaviors lower than the main load behaviors of the user. However, such events were likely normal electricity consumption behaviors created when the user was not at home that day. In addition, the direct use of low-value features could not effectively identify obvious and rapidly changing load behaviors (i.e., anomalies detected using high-level features). It is noteworthy that there is a minor performance difference between using low-level and high-level features in the experiment of User 05. To classify the point, we analyze the load profiles of User 05 clearly and find that the load profile of User 05 has a property compared with the other users; that is, User 05 has a more regular power consumption behavior than others. Therefore, this study can infer that the anomaly detection model using high-level features has a better performance than using low-level features under the situation in which the user has irregular power consumption behavior. Due to the randomness of power consumption behaviors in most users, the performance of anomaly detection using high-level features in general is better than using low-level features.

5. Conclusions

This study trains an autoencoder model and uses the network model to extract the low-level features (96-dimension features) of the residential power consumption data to be the high-level features (two-dimensional features) for improving the performance of abnormal power consumption behavior detection models for residential users. The experiments are implemented to prove that the anomaly detection model using the high-level features is more performance than the model using the low-level features in terms of identifying the abnormal power consumption behaviors of residential users. If the proposed technology can be integrated into the home energy management system (HEMS), HEMS can provide the appropriate energy-saving suggestions at a suitable timing due to the more accurate rate of anomaly behavior detection.

However, because the study adopted an unsupervised learning method to establish the anomaly detection mechanism, its explanatory power of abnormal electricity consumption behaviors remains insufficient. In addition, this study only uses visualization analysis to evaluate the performance of the anomaly detection models, and only the basic autoencoder model is used and evaluated. Therefore, how to improve the explanatory power of the output results of the anomaly detection model, how to find the explore quantitative indicators and methods that can more accurately compare the performances between the models trained using high-level and low-level features, and further using the more advanced deep learning model to improve the performance of anomaly detection models will be our future work.

Author Contributions: Conceptualization, C.-W.T. and K.-C.C.; methodology, C.-W.T., J.L. and C.-W.Y.; investigation, H.-Y.H.; formal analysis, C.-W.T. and Y.-C.C.; writing—original draft, C.-W.T. and K.-C.C.; writing—review and editing, J.L.; and project administration, Y.-C.C. All authors have read and agreed to the published version of the manuscript.

Funding: This research was partially supported by the Ministry of Science and Technology, Taiwan, R.O.C. (Grant Nos. MOST 110-2221-E-143-003, MOST 110-2221-E-259-001, MOST 110-2221-E-143-004, MOST 110-2221-E-039-004, MOST 110-2222-E-005-006, MOST 110-2634-F-005-006, and MOST 111-2218-E-005-007-MBK), Bureau of Energy, Ministry of Economic Affairs, Taiwan (Grant No. 111-E0208), and China Medical University, Taiwan (Grant No. CMU110-S-21).

Conflicts of Interest: The authors declare no conflict of interest.

Appendix A

To clearly represent each anomaly power consumption event indicated by the models using the low-level and high-level features, the individual load profiles of each user are shown in this appendix. Here, the green curve also denotes the central value of the primary load behavior of users, the light green and grass green areas indicate the Q1–Q4 range of the corresponding electricity consumption feature, and the red curve is the abnormal load profile.

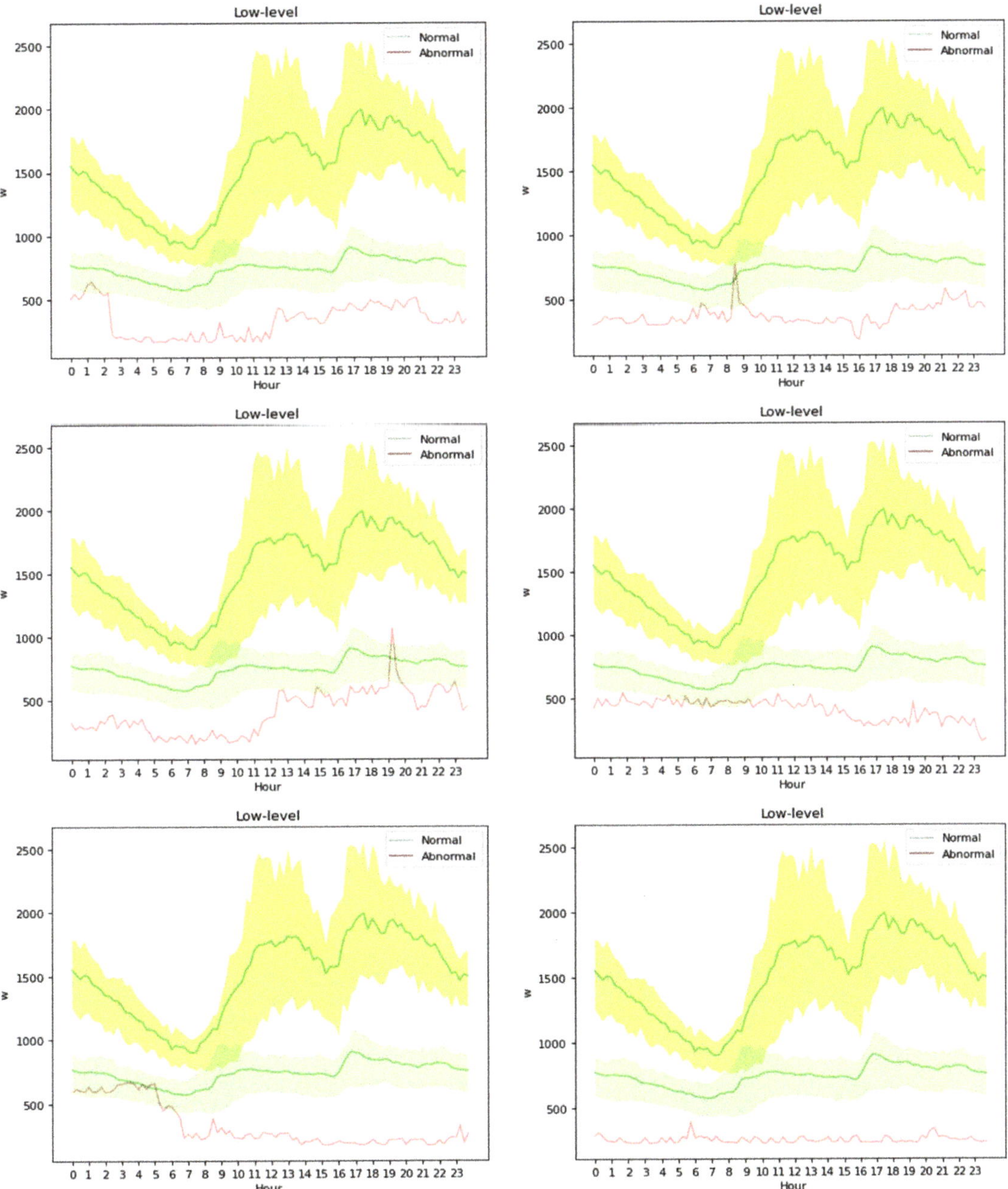

Figure A1. Load profiles of each abnormal electricity consumption events of User 01 using the low-level features.

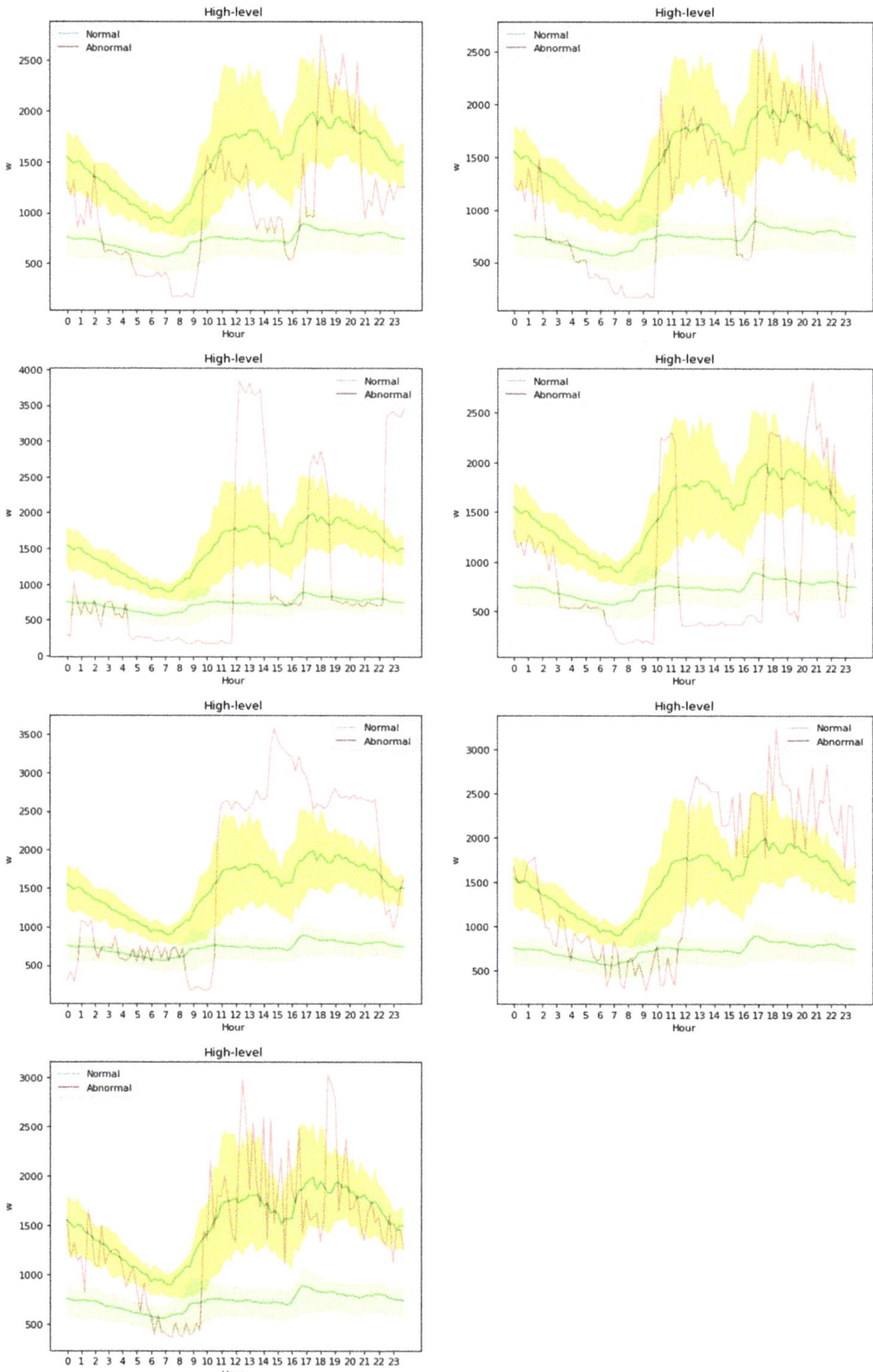

Figure A2. Load profiles of each abnormal electricity consumption events of User 01 using the high-level features.

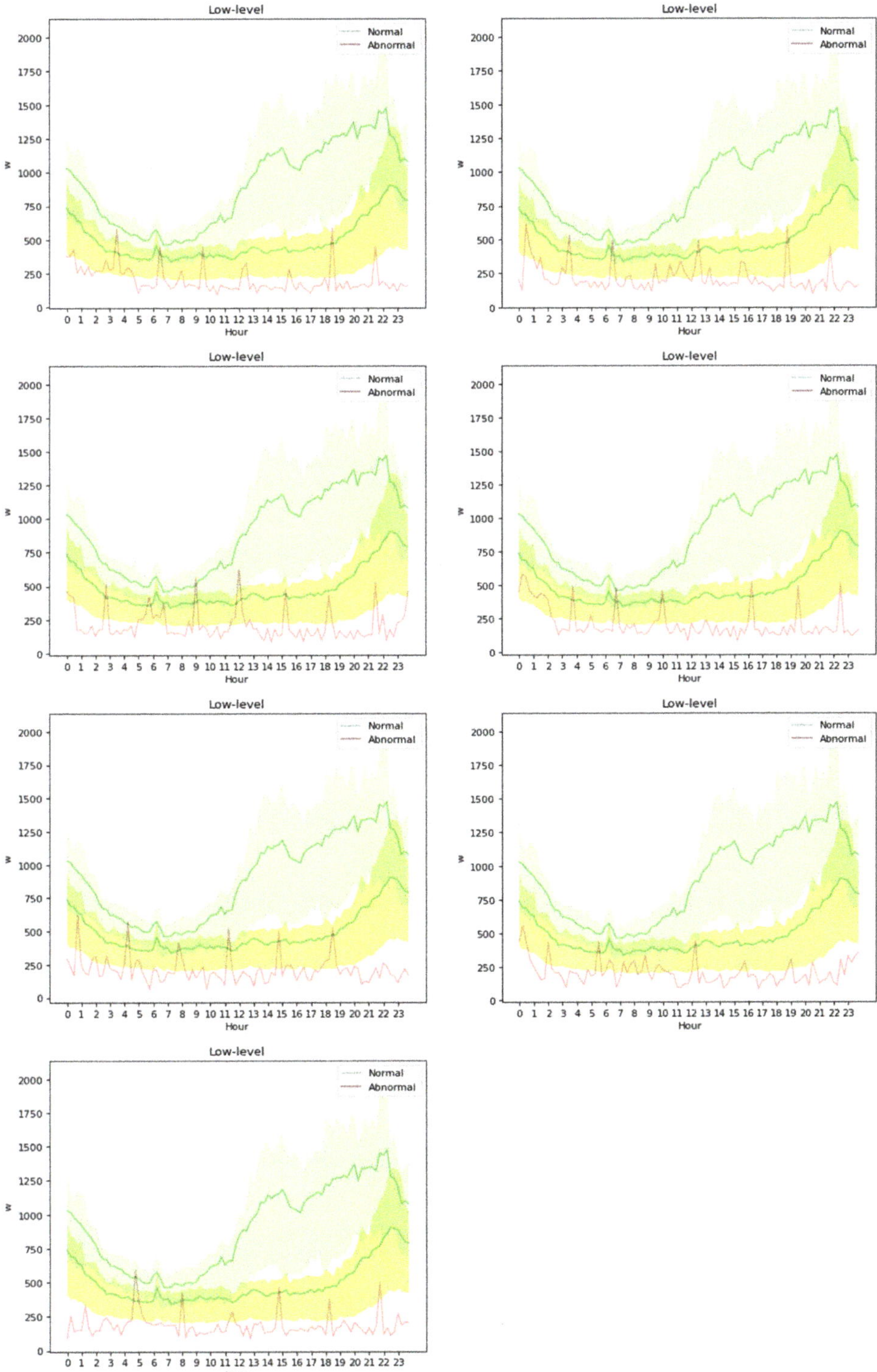

Figure A3. Load profiles of each abnormal electricity consumption events of User 02 using the low-level features.

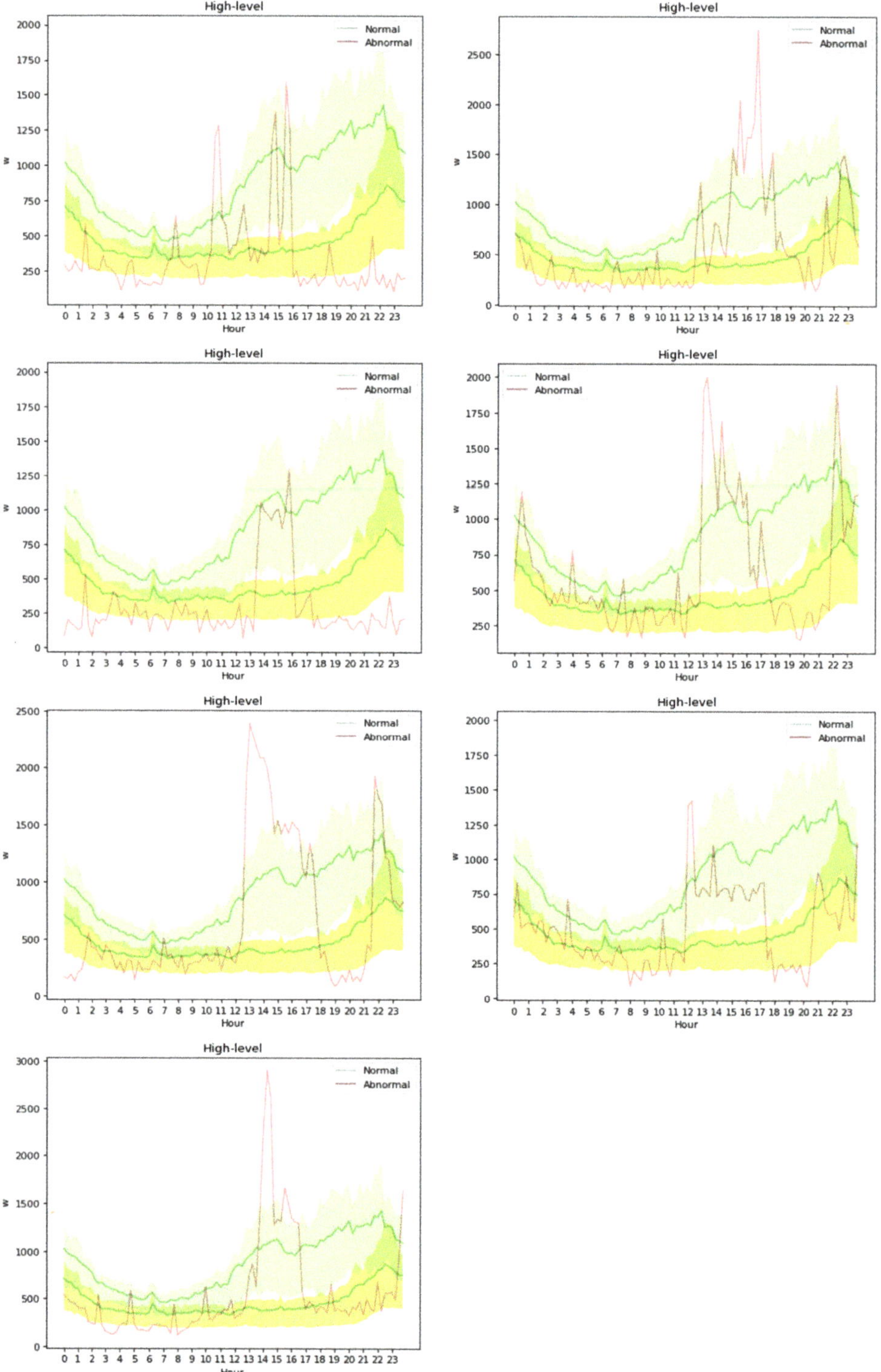

Figure A4. Load profiles of each abnormal electricity consumption events of User 02 using the high-level features.

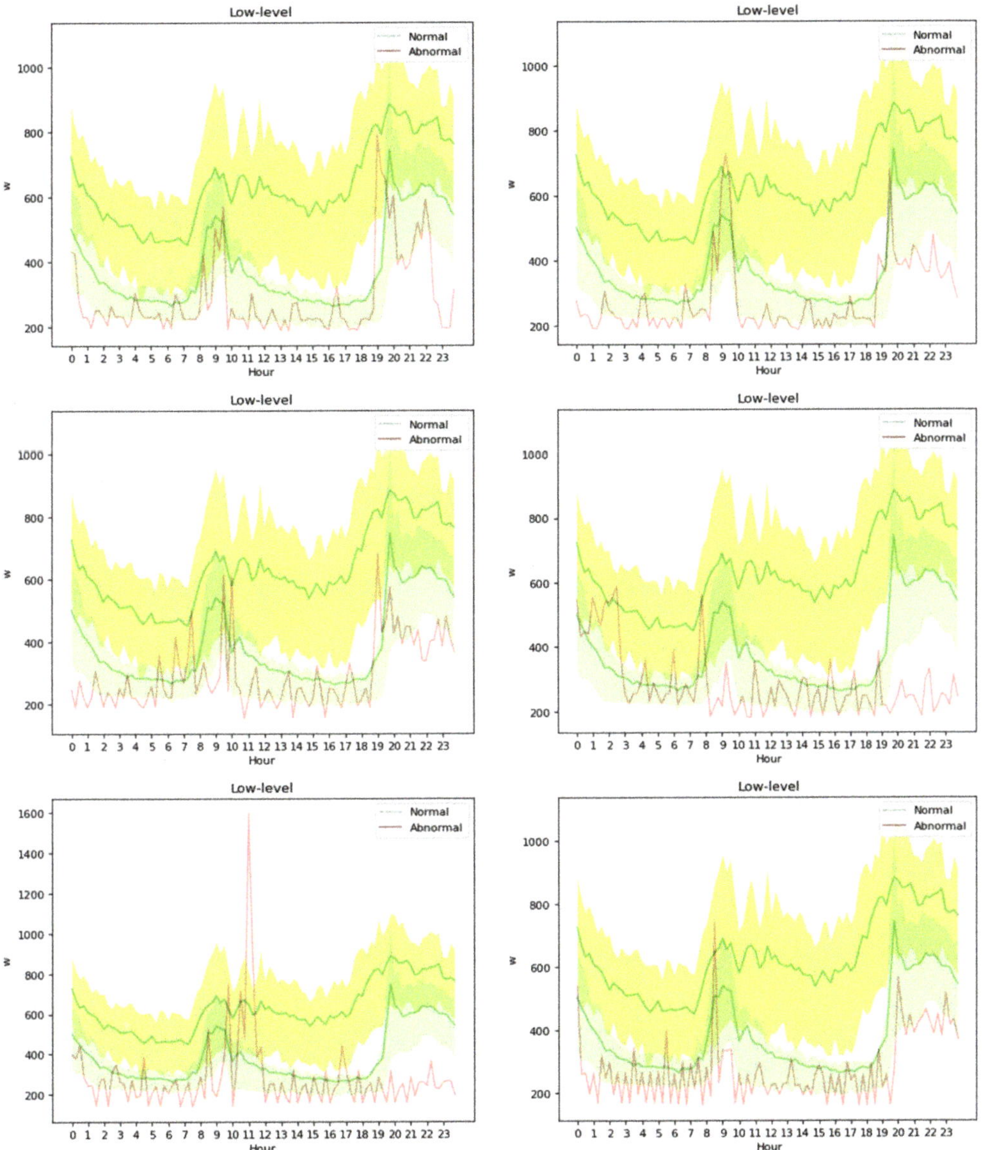

Figure A5. Load profiles of each abnormal electricity consumption events of User 03 using the low-level features.

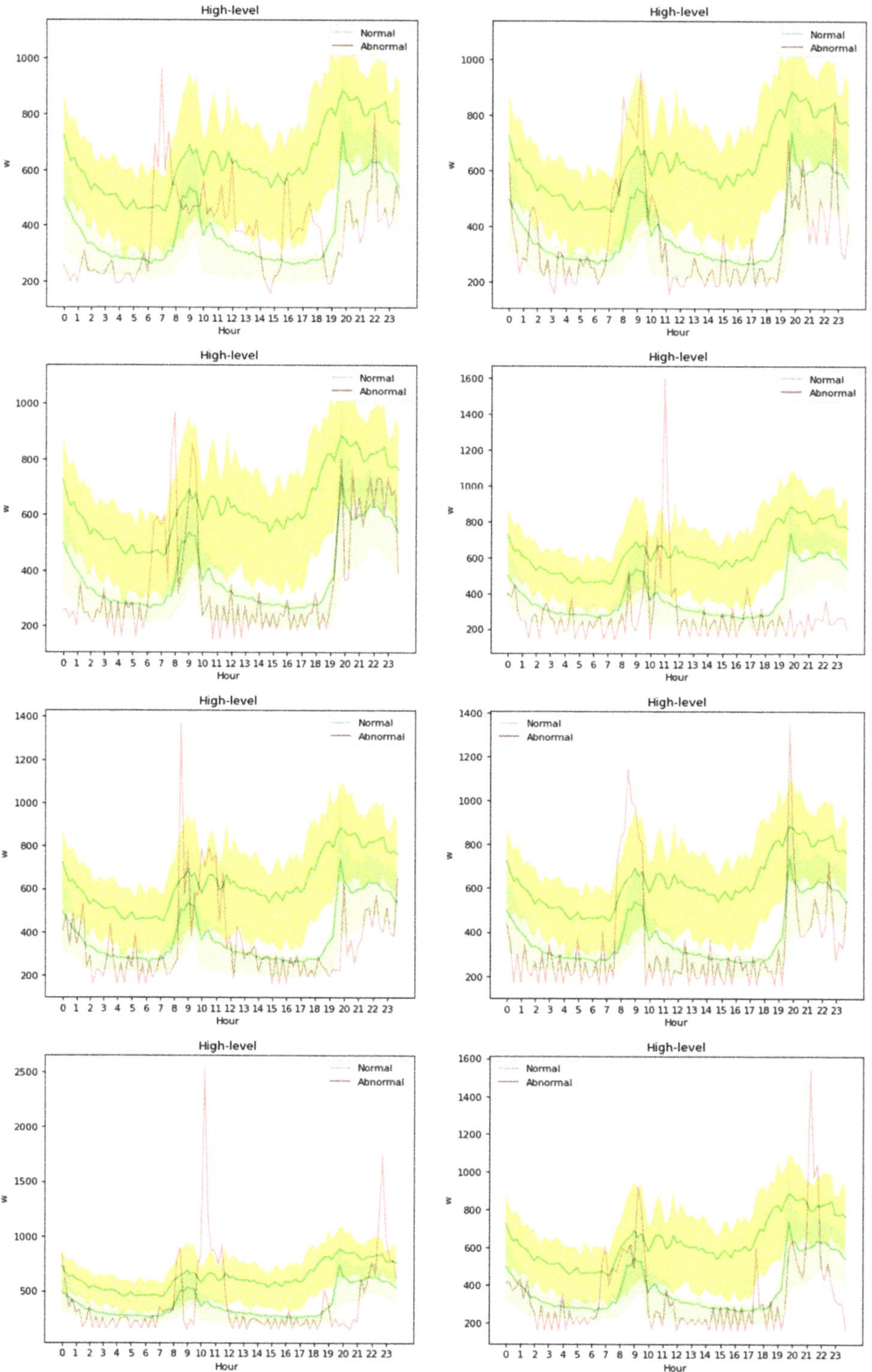

Figure A6. Load profiles of each abnormal electricity consumption events of User 03 using the high-level features.

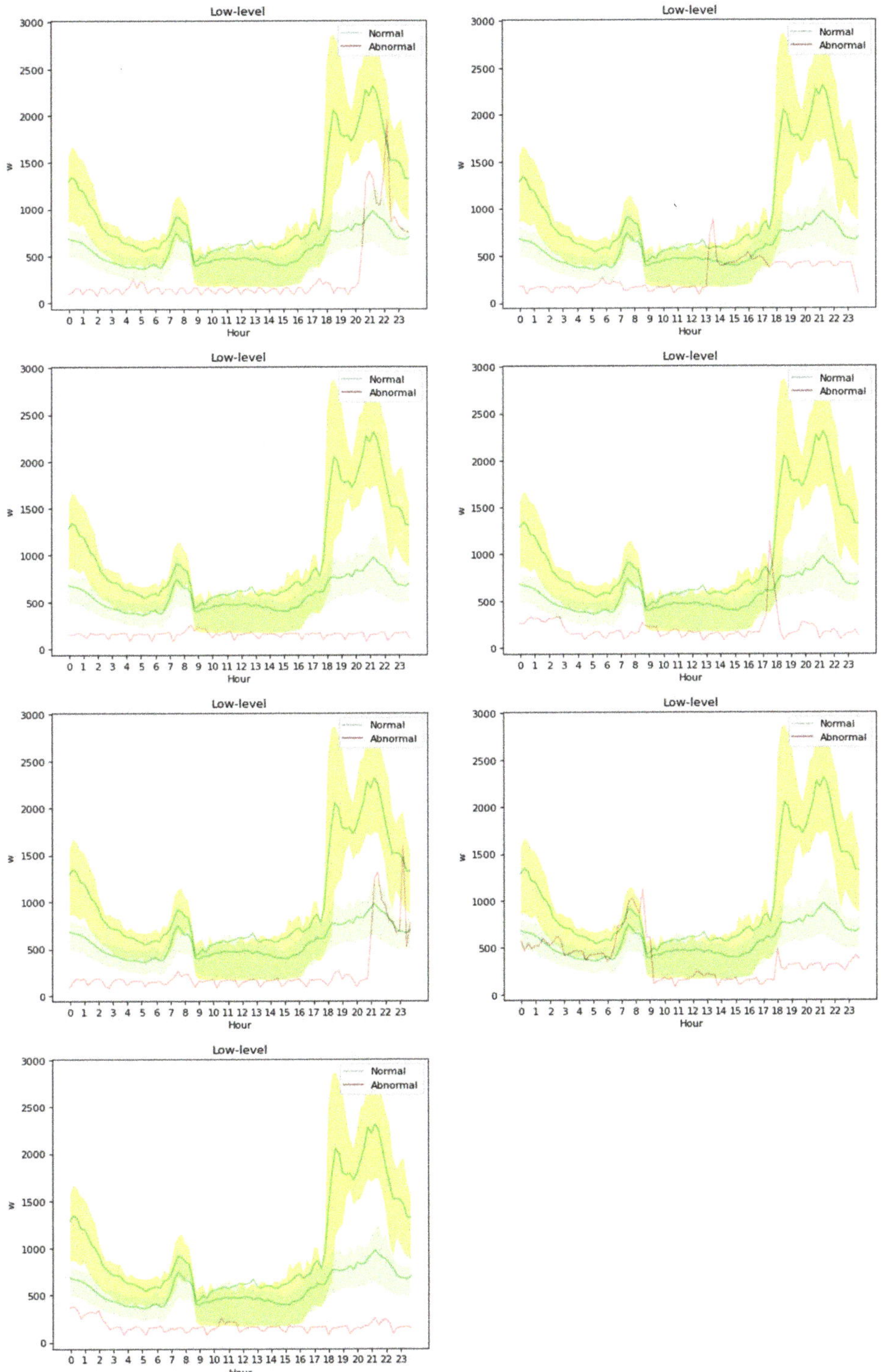

Figure A7. Load profiles of each abnormal electricity consumption events of User 04 using the low-level features.

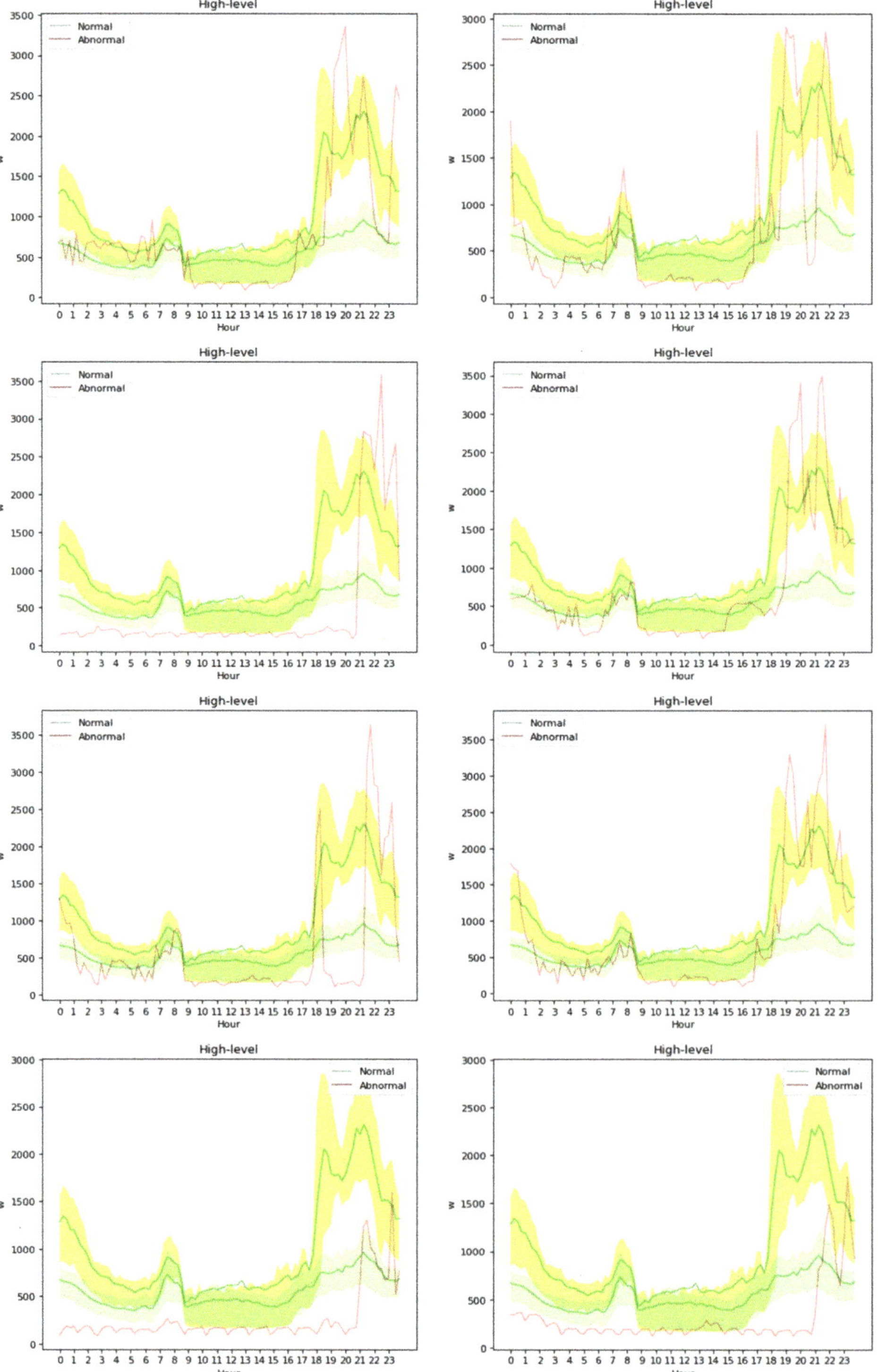

Figure A8. Load profiles of each abnormal electricity consumption events of User 04 using the high-level features.

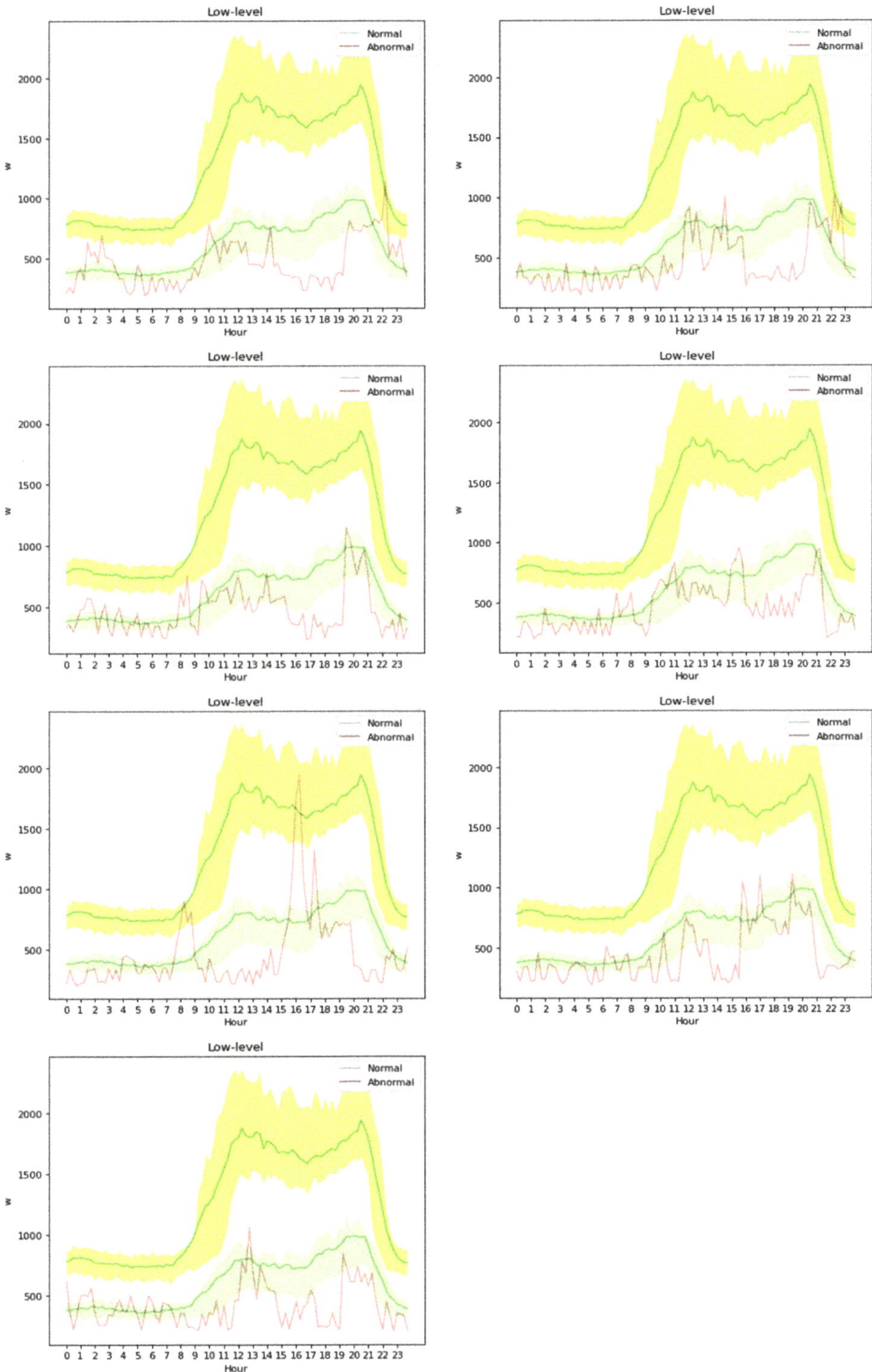

Figure A9. Load profiles of each abnormal electricity consumption events of User 05 using the low-level features.

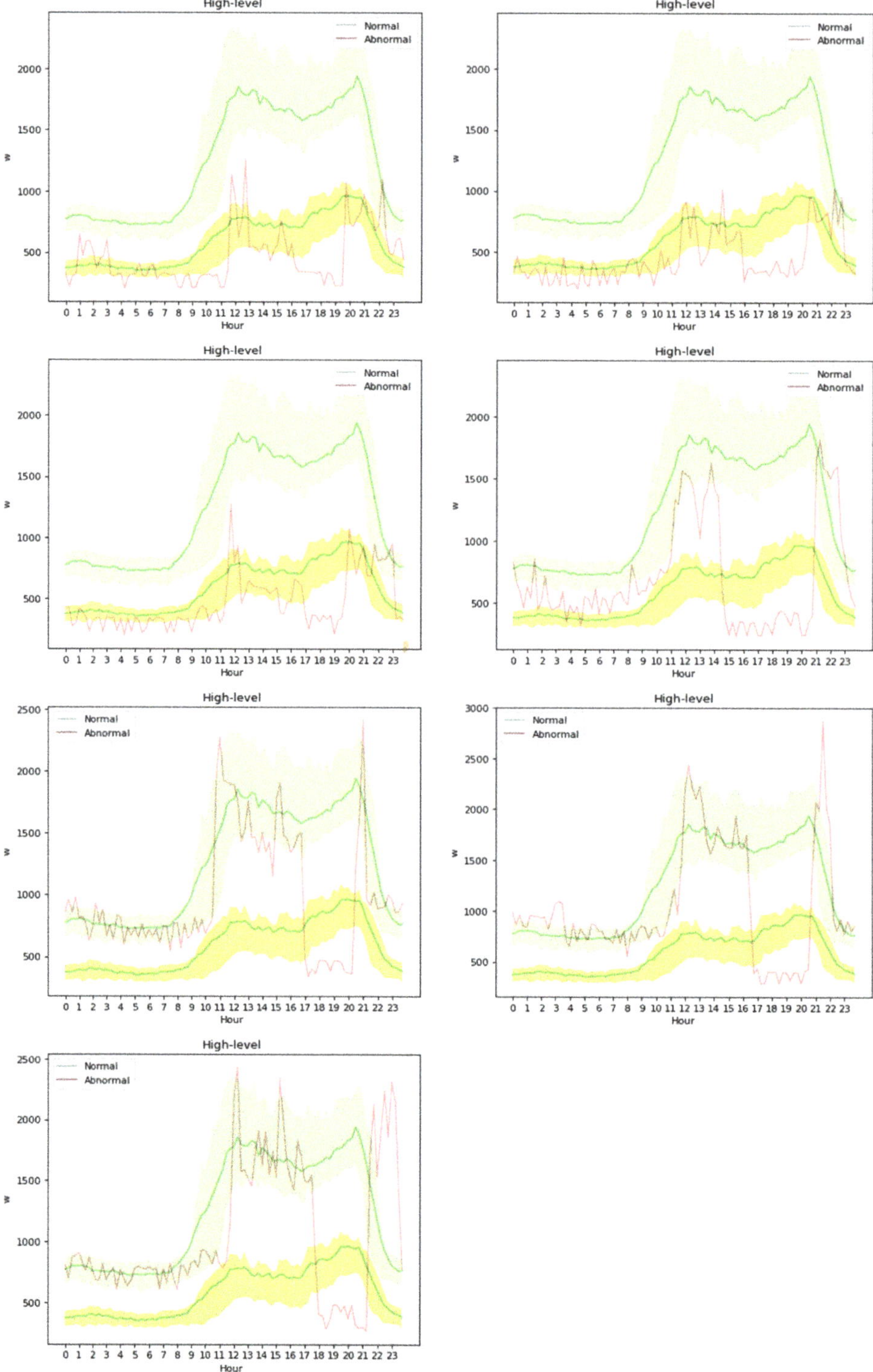

Figure A10. Load profiles of each abnormal electricity consumption events of User 05 using the high-level features.

References

1. Koolen, D.; Sadat-Razavi, N.; Ketter, W. Machine Learning for Identifying Demand Patterns of Home Energy Management Systems with Dynamic Electricity Pricing. *Appl. Sci.* **2017**, *7*, 1160. [CrossRef]
2. Seyedzadeh, S.; Rahimian, F.; Glesk, I.; Roper, M. Machine learning for estimation of building energy consumption and performance: A review. *Vis. Eng.* **2018**, *6*, 5. [CrossRef]
3. Al Tarhuni, B.; Naji, A.; Brodrick, P.G.; Hallinan, K.P.; Brecha, R.J.; Yao, Z. Large scale residential energy efficiency prioritization enabled by machine learning. *Energy Effic.* **2019**, *12*, 2055–2078. [CrossRef]
4. Yu, L.; Xie, W.; Xie, D.; Zou, Y.; Zhang, D.; Sun, Z.; Jiang, T. Deep reinforcement learning for smart home energy management. *IEEE Internet Things J.* **2019**, *7*, 2751–2762. [CrossRef]
5. Wang, A.; Lam, J.C.K.; Song, S.; Li, V.O.K.; Guo, P. Can smart energy information interventions help householders save electricity? a svr machine learning approach. *Environ. Sci. Policy* **2020**, *112*, 381–393. [CrossRef]
6. Machorro-Cano, I.; Alor-Hernández, G.; Paredes-Valverde, M.A.; Rodríguez-Mazahua, L.; Sánchez-Cervantes, J.L.; Olmedo-Aguirre, J.O. HEMS-IoT: A big data and machine learning-based smart home system for energy saving. *Energies* **2020**, *13*, 1097. [CrossRef]
7. Kumar, M.; Zhang, X.; Liu, L.; Wang, Y.; Shi, W. Energy-efficient machine learning on the edges. In Proceedings of the 2020 IEEE International Parallel and Distributed Processing Symposium Workshops (IPDPSW), New Orleans, LA, USA, 18–22 May 2020; pp. 912–921.
8. Naji, A.; Al Tarhuni, B.; Choi, J.; Alshatshati, S.; Ajena, S. Toward cost-effective residential energy reduction and community impacts: A data-based machine learning approach. *Energy AI* **2021**, *4*, 100068. [CrossRef]
9. Aslam, M.S.; Ghazal, T.M.; Fatima, A.; Said, R.A.; Abbas, S.; Khan, M.A.; Siddiqui, S.Y.; Ahmad, M. Energy-Efficiency Model for Residential Buildings Using Supervised Machine Learning Algorithm. *Intell. Autom. Soft Comput.* **2021**, *30*, 881–888. [CrossRef]
10. Revati, G.; Hozefa, J.; Shadab, S.; Sheikh, A.; Wagh, S.R.; Singh, N.M. Smart building energy management: Load profile prediction using machine learning. In Proceedings of the 2021 29th Mediterranean Conference on Control and Automation (MED), Bari, Italy, 22–25 June 2021; pp. 380–385.
11. Anastasiadou, M.; Vítor, S.; Miguel, S.D. Machine Learning Techniques Focusing on the Energy Performance of Buildings: A Dimensions and Methods Analysis. *Buildings* **2021**, *28*, 12. [CrossRef]
12. Zekić-Sušac, M.; Mitrović, S.; Has, A. Machine learning based system for managing energy efficiency of public sector as an approach towards smart cities. *Int. J. Inf. Manag.* **2020**, *58*, 102074. [CrossRef]
13. Shapi, M.K.M.; Ramli, N.A.; Awalin, L.J. Energy consumption prediction by using machine learning for smart building: Case study in Malaysia. *Dev. Built Environ.* **2020**, *5*, 100037. [CrossRef]
14. Chandola, V.; Banerjee, A.; Kumar, V. Anomaly detection: A survey. *ACM Comput. Surv.* **2009**, *41*, 1–58. [CrossRef]
15. Wang, X.; Ahn, S.-H. Real-time prediction and anomaly detection of electrical load in a residential community. *Appl. Energy* **2019**, *259*, 114145. [CrossRef]
16. Zhang, Y.; Chen, W.W.; Black, J. Anomaly Detection in Premise Energy Consumption Data. In Proceedings of the 2011 IEEE Power and Energy Society General Meeting, Detroit, MI, USA, 24–28 July 2011.
17. Chou, J.S.; Telaga, A.S. Real-time detection of anomalous power consumption. *Renew. Sustain. Energy Rev.* **2014**, *33*, 400–411. [CrossRef]
18. Luo, J.; Hong, T.; Yue, M. Real-time anomaly detection for very short-term load forecasting. *J. Mod. Power Syst. Clean Energy* **2018**, *6*, 235–243. [CrossRef]
19. Fenza, G.; Gallo, M.; Loia, V. Drift-aware methodology for anomaly detection in smart grid. *IEEE Access* **2019**, *7*, 9645–9657. [CrossRef]
20. Inayah, N.; Wijaya, M.Y.; Fitriyati, N. Energy Saving Potential Prediction and Anomaly Detection in College Buildings; In Proceedings of the International Conference on Mathematics and Islam (ICMIs 2018), Mataram, Indonesia, 3–5 August 2020.
21. Zhang, Y.; Lin, F.; Wang, K. Robustness of short-term wind power forecasting against false data injection attacks. *Energies* **2020**, *13*, 3780. [CrossRef]
22. Jokar, P.; Arianpoo, N.; Leung, V.C.M. Electricity theft detection in AMI using customers' consumption patterns. *IEEE Trans. Smart Grid* **2016**, *7*, 216–226. [CrossRef]
23. Pinceti, A.; Sankar, L.; Kosut, O. Load Redistribution Attack Detection Using Machine Learning: A Data-driven Approach. In Proceedings of the 2018 IEEE Power & Energy Society General Meeting (PESGM), Portland, OR, USA, 5–10 August 2018.
24. Fang, Z.; Cheng, Q.; Mou, L.; Qin, H.; Zhou, H.; Cao, J. Abnormal electricity consumption detection based on ensemble learning. In Proceedings of the 2019 9th International Conference on Information Science and Technology (ICIST), Inner Mongolia, China, 2–5 August 2019; pp. 175–182.
25. Wang, X.; Yang, I.; Ahn, S.-H. Sample Efficient Home Power Anomaly Detection in Real Time Using Semi-Supervised Learning. *IEEE Access* **2019**, *7*, 139712–139725. [CrossRef]
26. Welikala, S.; Dinesh, C.; Ekanayake, M.P.B.; Godaliyadda, R.I.; Ekanayake, J. Incorporating Appliance Usage Patterns for Non-Intrusive Load Monitoring and Load Forecasting. *IEEE Trans. Smart Grid* **2017**, *10*, 448–461. [CrossRef]
27. Hosseini, S.S.; Agbossou, K.; Kelouwani, S.; Cardenas, A.; Henao, N. A Practical Approach to Residential Appliances on-Line Anomaly Detection: A Case Study of Standard and Smart Refrigerators. *IEEE Access* **2020**, *8*, 57905–57922. [CrossRef]

28. Fan, C.; Xiao, F.; Zhao, Y.; Wang, J.Y. Analytical investigation of autoencoder-based methods for unsupervised anomaly detection in building energy data. *Appl. Energy* **2018**, *211*, 1123–1135. [CrossRef]
29. Pereira, J.; Margarida, S. Unsupervised Anomaly Detection in Energy Time Series Data Using Variational Recurrent Autoencoders with Attention. In Proceedings of the 2018 17th IEEE International Conference on Machine Learning and Applications (ICMLA), Orlando, FL, USA, 17–20 December 2018.
30. Weinberg, D. Our Machines Now Have Knowledge We Will Never Understand. Available online: https://www.wired.com/story/our-machines-now-have-knowledge-well-never-understand/ (accessed on 25 March 2022).
31. Janetzko, H.; Stoffel, F.; Mittelstädt, S.; Keim, D.A. Anomaly detection for visual analytics of power consumption data. *Comput. Graph.* **2014**, *38*, 27–37. [CrossRef]
32. Naganathan, H.; Chong, W.K.; Huang, Z.; Cheng, Y. A Non-stationary Analysis Using Ensemble Empirical Mode Decomposition to Detect Anomalies in Building Energy Consumption. *Procedia Eng.* **2016**, *145*, 1059–1065. [CrossRef]
33. Cabrera, D.F.M.; Zareipour, H. Data association mining for identifying lighting energy waste patterns in educational institutes. *Energy Build.* **2013**, *62*, 210–216. [CrossRef]
34. Li, Z.; Zareipour, H. Abnormal Energy Consumption Analysis Based on Big Data Mining Technology. In Proceedings of the 2020 Asia Energy and Electrical Engineering Symposium (AEEES), Chengdu, China,, 29–31 May 2020.
35. Rumelhart, D.E.; Hinton, G.E.; Williams, R.J. *Parallel Distributed Processing: Explorations in the Microstructure of Cognition, Chap. Learning Internal Representations by Error Propagation*; MIT Press: Cambridge, MA, USA, 1986; Volume 1, pp. 318–362. Available online: http://dl.acm.org/citation.cfm?id=104279.104293 (accessed on 25 March 2022).
36. Baldi, P. Autoencoders, unsupervised learning, and deep architectures. In Proceedings of the ICML Workshop on Unsupervised and Transfer Learning, JMLR Workshop and Conference Proceedings, Edinburgh, Scotland, 26 June–1 July 2012.
37. Schölkopf, B.; Williamson, R.; Smola, A.; Shawe-Taylor, J.; Platt, J. Support Vector Method for Novelty Detection. In Proceedings of the 12th International Conference on Neural Information Processing Systems, Denver, CO, USA, 29 November–4 December 1999.
38. Li, K.L.; Huang, H.K.; Tian, S.F.; Xu, W. Improving one-class SVM for anomaly detection. In Proceedings of the 2003 International Conference on Machine Learning and Cybernetics (IEEE Cat. No. 03EX693), Xi'an, China, 2–5 November 2003; Volume 5, pp. 3077–3081.
39. Wang, Y.; Johnny, W.; Andrew, M. Anomaly intrusion detection using one class SVM. In Proceedings of the Fifth Annual IEEE SMC Information Assurance Workshop, IEEE, New York, NY, USA, 10–11 June 2004.

Article

A Cloud-Edge-Smart IoT Architecture for Speeding Up the Deployment of Neural Network Models with Transfer Learning Techniques

Tz-Heng Hsu [1,*], Zhi-Hao Wang [2] and Aaron Raymond See [3]

[1] Department of Computer Science and Information Engineering, Southern Taiwan University of Science and Technology, Tainan 710301, Taiwan

[2] Department of Information Management, Southern Taiwan University of Science and Technology, Tainan 710301, Taiwan; zhwang@stust.edu.tw

[3] Department of Electrical Engineering, Southern Taiwan University of Science and Technology, Tainan 710301, Taiwan; aaronsee@stust.edu.tw

* Correspondence: hsuth@mail.stust.edu.tw; Tel.: +886-6-253-3131 (ext. 3226)

Abstract: Existing edge computing architectures do not support the updating of neural network models, nor are they optimized for storing, updating, and transmitting different neural network models to a large number of IoT devices. In this paper, a cloud-edge smart IoT architecture for speeding up the deployment of neural network models with transfer learning techniques is proposed. A new model deployment and update mechanism based on the share weight characteristic of transfer learning is proposed to address the model deployment issues associated with the significant number of IoT devices. The proposed mechanism compares the feature weight and parameter difference between the old and new models whenever a new model is trained. With the proposed mechanism, the neural network model can be updated on IoT devices with just a small quantity of data sent. Utilizing the proposed collaborative edge computing platform, we demonstrate a significant reduction in network bandwidth transmission and an improved deployment speed of neural network models. Subsequently, the service quality of smart IoT applications can be enhanced.

Keywords: deep learning; transfer learning; lightweight neural network; edge computing

Citation: Hsu, T.-H.; Wang, Z.-H.; See, A.R. A Cloud-Edge-Smart IoT Architecture for Speeding up the Deployment of Neural Network Models with Transfer Learning Techniques. *Electronics* **2022**, *11*, 2255. https://doi.org/10.3390/electronics11142255

Academic Editors: Antoni Morell and Juan-Carlos Cano

Received: 1 June 2022
Accepted: 15 July 2022
Published: 19 July 2022

Publisher's Note: MDPI stays neutral with regard to jurisdictional claims in published maps and institutional affiliations.

1. Introduction

With the rapid development of mobile broadband network communications and artificial intelligence technology, smart video networking services are becoming more and more popular. In order to satisfy the needs of large amounts of data and low transmission delays for video networking devices, edge computing architectures have emerged. The trend of artificial intelligence computing is moving from cloud computing to edge computing that uses a decentralized architecture to shorten the delay of network transmission and accelerate the processing speed of real-time computation. Different computing tasks are processed hierarchically and computing is completed close to the data source or client side to shorten network transmission delays and quickly obtain data analysis results. Edge computing speeds up the response speed of application services through the concept of computing layering and provides a better user experience. To meet the requirements of artificial intelligence (AI) services, the trend of AI computing is moving from cloud platforms to distributed edge computing. Existing edge computing frameworks, such as fog computing, cloudlet, and mobile edge computing, still have many shortcomings [1].

The need to deploy machine learning models on edge devices is growing rapidly. Edge AI allows inference to be run locally without connecting to the cloud, reducing the data transmission time and making the inference process of neural network models more efficient. Many vendors offer AI edge computing deployment platforms for IoT devices, e.g., Microsoft Azure IoT Edge [2], AWS Greengrass [3] Google's Cloud IoT Edge [4], and IBM Watson IoT Platform Edge [5]. Microsoft Azure IoT Edge is built upon Azure IoT Hub that offers a centralized method of managing Azure IoT edge devices and deploying the neural network models to the edge devices. Users that prefer to perform AI tasks at the edge rather than in the cloud can use this service, in which IoT devices can spend less time sending data to the cloud server by shifting AI tasks to the edge. The workflow of deploying machine learning (ML) models on the Azure IoT Edge platform [6] is as follows: first, create Azure resources; second, configure Edge device; third, build the ML model into the docker image; fourth, deploy the ML model to the IoT Edge; fifth, test the ML module; and finally, tear down resources. In the third step, the user develops an ML model, builds it into a docker image, and registers it into an Azure Container Registry (ACR). In the fourth step, the user deploys modules (running containers from docker images registered in ACR) to the IoT edge device. In the Microsoft Azure IoT Edge architecture, when a neural network model is retrained, the entire neural network model file must be repackaged into a new docker image and then deployed to IoT devices, which is time consuming and requires a large data size to encapsulate the neural network model into the docker image file.

Larger and more diverse data are being gathered as various sensors are placed in mobile phones, vehicles, and buildings. How the diversity and large size of sensor data integrates into artificial intelligence (AI) solutions is an interesting topic of research [7]. Hence, how to provide a better service experience for users has become an important issue for intelligent AI services.

Traditional machine learning methods include supervised learning, unsupervised learning, and semi-supervised learning. One of the problems with the main visual neural network model of supervised learning is that training a new model from scratch requires a lot of data. If there are not enough training data, the trained neural networks are often prone to overfitting. However, in many application environments, due to various restrictions, it is hard to collect enough training samples for the neural network to effectively converge. Transfer learning technology can solve the overfitting problem of insufficient image samples. It usually starts by initializing a new neural network that retrieves the trained neural layer and weight information in advance from the publicly available high-performance neural network models, and then performs local weight updates. The pre-trained neural network can provide the trained feature weights as an advanced feature extractor. Transfer learning can also be applied to the training of lightweight neural network models, which can be implemented in hardware accelerators; the last layer of the target domain neural network can be quickly retrained and inferred on the hardware of the networked devices [8,9].

Transfer learning is the process of using feature representations from a pre-trained model to avoid having to train a new model from the ground up. Pre-trained models are typically trained on large datasets, which are common benchmarks in the fields of computer vision and natural language processing [10]. Transfer learning has a number of advantages when it comes to the construction of machine learning models. The weights calculated by the pre-trained models can be used as part of the training process for a new model. Generalized information can be transferred across the two neural network models if they are designed to accomplish similar tasks. When pre-trained models are used in a new model, the required computing resources and training time can be reduced [11]. Meanwhile, in the case of only having a tiny training dataset, transfer learning technologies can utilize the weights from the pre-trained models to train the new model and solve new challenges. Transfer learning has been effectively used in a variety of machine learning applications, including text sentiment classification [12] and image classification [13,14].

The main purpose of this paper is to develop a collaborative edge computing platform that aims to solve the problems of (1) deploying large numbers of neural network models to IoT devices and (2) reducing the computing resource requirements of cloud servers. With the advance of deep learning algorithms, the file size of the trained model is growing larger and more parameterized in order to increase the prediction accuracy, which creates a deployment issue. When a large number of edge AI computing devices update the new models, the amount of data transmission required is considerable. For neural network models trained with transfer learning technologies, these models will share some of the same weights and parameters. Based on this characteristic, we propose a new model deployment and updating mechanism for solving the problems of deploying large numbers of neural network models to IoT devices. Whenever a new model is trained, the proposed mechanism will compare the feature weight and parameter difference between the old and new models. After the new neural network model is trained, the feature weights are encapsulated into a neural network model patch file after processing through a difference comparison (diff). By dispatching the neural network model patch file to all edge IoT devices for new model updating, only a small amount of data are required to be transmitted to the edge device, and the entire neural network model can be updated. As a result, the amount of data transmission can be significantly decreased for updating edge devices that already have the same shared pre-trained models. A collaborative model publish system is also proposed to assist the delivery and update of neural network models, reducing backbone bandwidth requirements. In order to evaluate the performance of the proposed model deployment and update mechanism, two different neural network models are used to evaluate the performance of the proposed system. The experimental results show that the proposed deployment and update mechanism can significantly lower the bandwidth requirement in model data transmission volume. Meanwhile, the computing resource requirements of cloud servers can be reduced with the proposed collaborative edge computing platform.

2. Related Works

Commercial cloud-based deep learning systems, such as Amazon Rekognition, usually require users to upload their personal data to a remote server for high-quality inference processing. However, uploading all data to a remote location may result in massive data transfers [15] and potential privacy risks [16]. Edge computing is a distributed computing architecture that moves the operations of applications, data, and services from the central node of the network to the edge nodes on the network logic for processing. Edge computing includes the following elements: (1) Proximity is in the Edge: Communication between edge nodes is faster and more efficient than communication with remote servers. (2) Intelligence is in the Edge: With the continuous enhancement of the computing power of sensors and smart video networking devices, edge nodes can make autonomous decisions and make immediate responses to the sensed data. (3) Trust is in the Edge: Much sensitive data are usually stored in personal devices (edge nodes). Therefore, trust relationships and sensitive data must also be managed manually at the Edge. (4) Control is in the Edge: Computing, data synchronization, or storage can be selectively allocated or delegated to other nodes or cores, and controlled by edge devices. (5) Humans are in the Edge: Human-centered design should place humans in a control loop, allowing users to control their data. Edge computing decomposes large-scale services that were originally handled by central nodes, cuts them into smaller and easier-to-manage parts, and distributes them to edge nodes for processing. As the edge node is closer to the user terminal device, it can speed up the processing and transmission of data and reduce transmission delay [17].

There are three main types of edge computing implementation architecture, which can be classified into fog computing, cloudlets, and mobile edge computing (MEC) [18]. Fog computing is a decentralized computing architecture based on fog computing nodes. Fog nodes consist of multiple elements, including routers, switches, network access points, IoT gateways, and set-top boxes. These nodes can be deployed anywhere between network architectures. Because these nodes are close to the edge of the network, fog computing can provide good real-time transmission quality. The abstract layer of fog computing can mask the differences between the devices, unify the resources of the devices, and form a resource pool that can be used by the upper layer. The edge network is very close to the end user, and the sensing data will be calculated directly at the fog computing nodes without uploading to the cloud computing center. Fog computing makes full use of a large number of smart devices located at the edge of the network. Although the computing resources of individual devices are limited, a large number of devices can play a significant role in a centralized manner. Fog computing can use various heterogeneous networks, such as wired, wireless, and mobile networks, to connect different network devices, which can solve the network delay problem of cloud computing [19]. Scholars such as Beck introduced the classification and architecture topology of Mobile Edge Computing [20], as well as emerging technologies and applications of Edge content delivery and aggregation.

Many mobile devices now incorporate a large number of artificial intelligence application services, such as face recognition and voice translation. However, these mobile devices have very limited resources, such as limited battery power, network bandwidth, and computing storage capacity. Most artificial intelligence application services upload data to a cloud computing server, use the powerful computing and storage capabilities of the cloud computing center to process and store the data, and then send the computing results back to the mobile devices. However, with the development of the Internet of Things and the rapid growth of mobile devices, the transmission of large amounts of data to the cloud computing center can easily cause backbone network congestion, and network latency has become the bottleneck of cloud computing. Therefore, some scholars have proposed the concept of cloudlets. Cloudlets are resource-rich hosts or clusters of hosts. They are placed at the edge of the network and placed on the network closest to the mobile devices [21]. In this way, data can be sent to cloudlets for processing and return the results. At the same time, calculations that cannot be completed by the cloud nodes can be transferred to the cloud computing center for processing by the cloud server. Cloudlets have the following characteristics: (1) Proximity: cloudlets are very close to mobile devices and can be reached through one hop of the network. (2) Resource-Rich: compared with fog computing nodes, cloudlets are specially deployed computing or data storage nodes, and their computing and data storage capabilities are much higher than those of fog computing [21].

Edge computing architectures such as fog computing, cloudlets, and MEC still have much room for improvement in the implementation of artificial intelligence application services. The traditional edge computing architecture and collaborative algorithms do not support the training update of deep neural network models and the pre-trained network models; therefore, how to provide an efficient edge server that supports various neural network model updates has become a major issue for the successful deployment of large-scale artificial intelligence application services.

3. System Architecture

Figure 1 shows the proposed Cloud-Edge Smart IoT architecture and model publish system for speeding up the deployment of neural network models with transfer learning techniques. The proposed cloud AI server uses transfer learning technology to retrieve the pre-trained neural layers and the weight information and then updates the local weights of the newly trained neural network model. A model publish system is developed to help the neural network model delivery. After the new feature weights have been trained, the cloud AI server uses the proposed model publish system to encapsulate the new feature weights into a neural network model patch file, i.e., the diff part. Then, the model patch file is transmitted to the edge servers. When an edge server receives the new model patch file, the edge server stores the patch file into its cache storage space and then notifies its nearby smart edge IoT devices for neural network model updating. When the nearby smart edge IoT devices receive the notification, the IoT devices then send model update requests to download the new model patch file and update their local neural network models if the old models exist. In this way, it is no longer necessary to transmit big-data-volume neural network model files; the cloud server only needs to transmit small-data-size neural network model patch files to smart edge IoT devices with the help of edge servers. When an edge IoT device receives the patch file, the weights can be updated in combination with the existing models within the edge IoT device. With the help of the proposed collaborative edge computing platform, network bandwidth transmission requirements can be greatly reduced, and the deployment speed of neural network models can be improved. This section describes the proposed model publish system for speeding up the deployment of neural network models with transfer learning techniques.

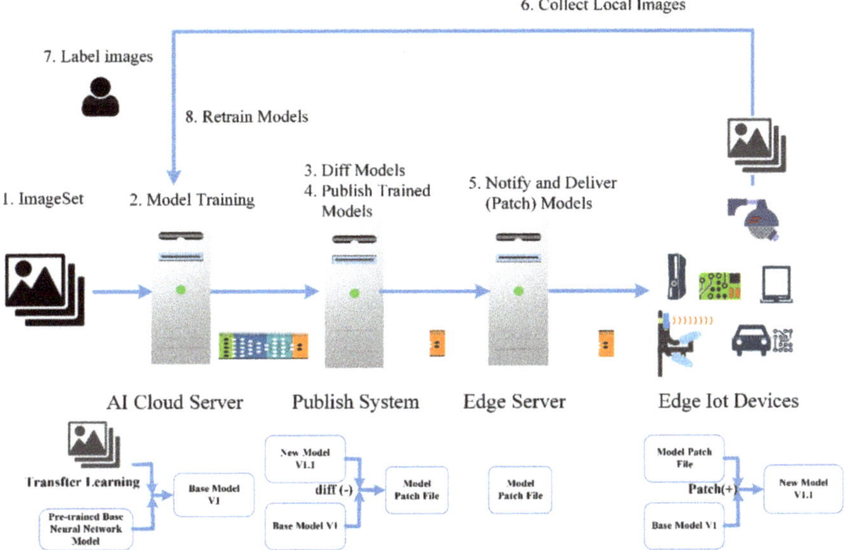

Figure 1. The proposed Cloud-Edge Smart IoT architecture and model publish system for speeding up the deployment of neural network models with transfer learning techniques.

3.1. The Model Publish System for Speeding Up the Deployment of Neural Network Models with Transfer Learning Techniques

Figure 2 illustrates the training mechanisms of deep neural networks (DNNs). One or more Hidden Layers (HLs) and Output Layer (Ols) are trained with the input dataset. Figure 2a represents the traditional machine learning (ML) mechanism, and various datasets are used to independently train separate models on given problems. Every trained model has a separate set of parameters, and no knowledge that can be shared between models. Subsequently, the entire neural network model must be retrained each time new data are collected, which takes a lot of training time and computing resources. Figure 2b represents the transfer learning mechanism; users can train new models using existing knowledge (features, weights, etc.) from pre-trained models and even solve problems such as having less data for the newer task. The most significant advantages of transfer learning are resource savings and increased efficiency while training new models [22]. Meanwhile, some weights and parameters could be shared by neural network models trained with transfer learning technologies. Based on this feature, we propose a new model deployment and updating mechanism to address the issues associated with deploying large numbers of neural network models to IoT devices.

In the proposed system architecture, once new data are collected and a neural network model needs to update its weights, the cloud AI servers use the same basic neural network as that on the edge IoT devices to perform weight update training through the transfer learning technologies to generate a new neural network model. After the new neural network model is trained, the feature weights that have been preprocessed and trained are encapsulated into a neural network model patch file after processing through a difference comparison (diff) with the help of the proposed model publish system. The model publish system will use the existing neural network model's weights to dispatch the neural network model patch file to all edge IoT devices for model updating through the help of edge servers.

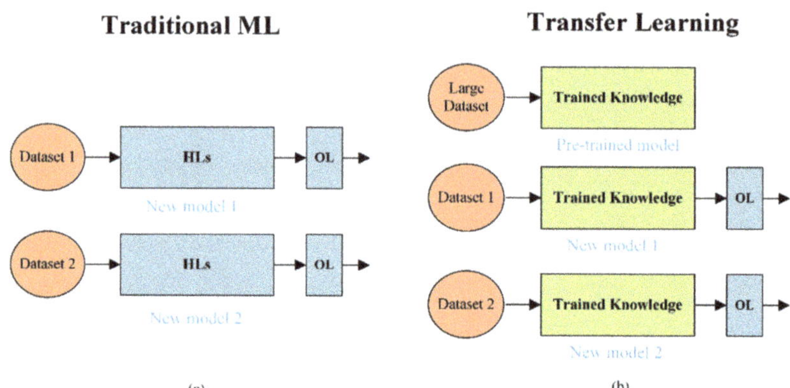

Figure 2. The training mechanisms of deep neural networks (DNNs): (**a**) traditional machine learning (ML) and (**b**) transfer learning.

Figure 3 shows the workflow chart of the proposed model diff/patch mechanism. When new data are collected, the cloud server trains the new neural network model with system configurations. If a model is suitable for use with transfer learning technologies, then the model is trained with a configured pre-trained model. Otherwise, the model is trained from scratch. Once a new model has been trained, the system checks to see if the previous model already exists for the same AI task. The system employs the diff tool to identify the differences between the old and the newly trained models if an old model is present. Then, a model patch file is created to contain all of the model's different pieces. The model patch file is delivered to IoT devices through the help of edge servers. If no old model exists in the cloud storage system, the new trained model file is delivered. Finally, all IoT devices update the existing old model with the new model patch file or have a whole new model file and perform the new AI task.

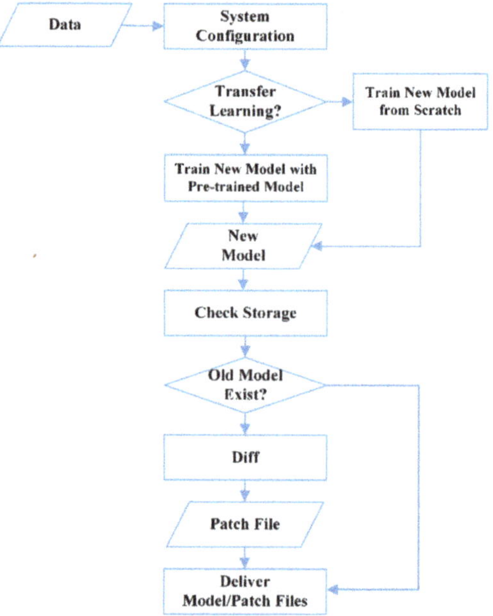

Figure 3. The workflow chart of the proposed model diff/patch mechanism.

3.2. The Cloud-Edge Smart IoT Architecture

This section explains the system model used in the proposed cloud-edge-Smart IoT architecture. Assume that there are N edge servers and each edge server has M child Smart IoT nodes in the cloud-edge Smart IoT network. Figure 4 shows the timing diagram of the proposed cloud-edge Smart IoT architecture, where TR_{upload} represents the time required by a user to upload the labeled data to the cloud AI server; TR_{train_model} represents the model training time required by the cloud AI server; T_{tm} represents the start time of model training on the cloud AI server; T_{dm} denotes the start time of the *diff* old model on the cloud AI server if an old model exists; TR_{diff_model} denotes the model *diff* time processed at the cloud AI server. Assume that an edge server's average available bandwidth for downloading a model from the cloud server is denoted as BW_{cloud_edge} and the average model size is denoted as $Model_{size}$. T_{ce} denotes the start time of dispatching a new model to an edge server. Let TR_{cloud_edge} denote the transmission time of dispatching a new model to the edge server, which is equal to:

$$TR_{cloud\_edge} = TR_{notify\_edge} + TR_{request\_edge} + \frac{Model_{size}}{BW_{cloud\_edge}} \tag{1}$$

where TR_{notify_edge} denotes that the cloud server notifies an edge server that a new model is trained completely, and $TR_{request\_edge}$ denotes that the edge server sends a new model update request to the cloud AI server. Assume that a Smart IoT node's average available bandwidth for downloading the model from the edge server is denoted as BW_{edge_IoT}, and T_{ei} denotes the start time of dispatching a new model from the edge server to a Smart IoT node. Let TR_{edge_IoT} denote the transmission time of dispatching a new model from the edge server to a Smart IoT node, which is equal to:

$$TR_{edge\_IoT} = TR_{notify\_IoT} + TR_{request\_IoT} + \frac{Model_{size}}{BW_{edge\_IoT}} \tag{2}$$

where TR_{notify_IoT} denotes that the edge server notifies a Smart IoT node that a new model is trained completely, and $TR_{request\_IoT}$ denotes that the Smart IoT node sends the new model update request to the corresponding edge server. T_{pi} denotes the start time of patching the new model in the Smart IoT node. The total model patch time in the Smart IoT node is denoted as TR_{patch_IoT}. In the proposed Cloud-Edge-Smart IoT architecture, the total time TR_{total} for updating a new model to $N*M$ smart edge IoT devices is equal to:

$$TR_{total} = TR_{upload} + TR_{train\_model} + TR_{diff\_model} + N * TR_{cloud\_edge} + N * M * \left(TR_{edge\_IoT} + TR_{patch\_IoT} \right) \tag{3}$$

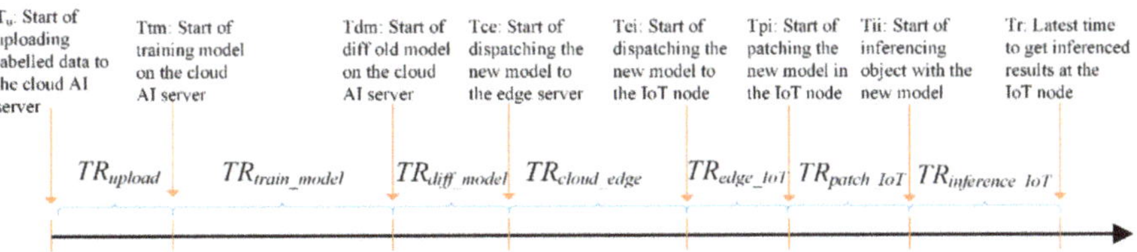

Figure 4. The timing diagram of the Cloud-Edge-SmartIot Architecture.

After the new model is patched completely, the edge IoT device can start the AI inference tasks. Let T_{ii} denote the start time of the inferencing object with the new patched model in the Smart IoT node. The total inferencing time in the Smart IoT node is denoted as $TR_{inference\_IoT}$; T_r denotes the latest time to obtain the inferenced result at the Smart IoT node.

4. System Evaluations

In the proposed collaborative edge computing platform, we use the Django framework [23] to implement the neural network model publish server and use the Celery framework [24] to perform the model *diff* task in the background; the Redis broker [25] server is used to store the Celery tasks. In Celery, task queues are used to distribute work across threads and machines. Dedicated worker processes constantly monitor task queues for new work to perform. Celery usually uses a broker, i.e., Redis broker, to help clients and workers talk to each other through messages. A Celery system can have multiple workers and brokers, allowing for high availability and horizontal scaling.

Using a publish/subscribe model, the MQTT protocol [26] provides a lightweight method of messaging. The MQTT protocol is appropriate for Internet of Things messaging, such as with low-power sensors and mobile devices. In the proposed collaborative edge computing platform, when a neural network model *diff* task is completed, the Mosquito MQTT broker [27] server is used to notify the edge servers to download and update the existing model. After the edge server downloads the model, it will notify the edge IoT devices to update the local neural network model through the Mosquito MQTT broker server. Each IoT device in the proposed platform comes with a pre-installed client system, which will automatically monitor the MQTT messages, update the AI model, and hand off control to the AI task for making inferences after the model is updated. Figure 5 shows the detailed workflow of the neural network model publish process in the proposed collaborative edge computing platform.

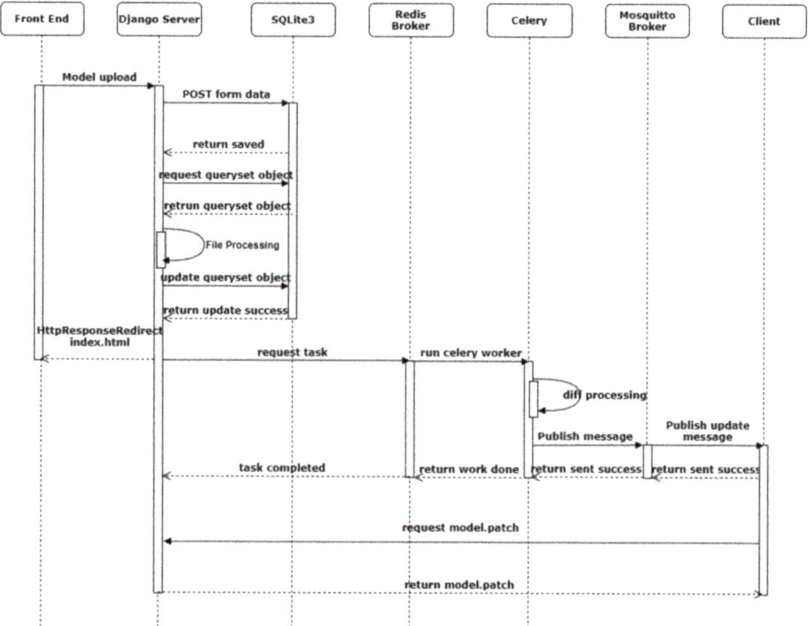

Figure 5. The detailed workflow of the neural network model publish process in the proposed collaborative edge computing platform.

In the experiment, a model publish server used in the experiment utilized an Intel(R) Core (TM) i9-9900K CPU @ 3.60 GHz, 64 GB RAM, a GeForce RTX 2080 Ti video card, a 512 GB SSD, and the operating system of Ubuntu 20.04.3 LTS. The experiment uses the Google Colab (Colaboratory) deep learning platform for model development with TensorFlow 2.5 and the Keras framework. The basic neural network test models used are MobileNet v2 [28] and VGG-19. MobileNet v2 enhances mobile model performance across many benchmarks and model sizes, which is built on an inverted residual structure. In the intermediate expansion layer, MobileNet v2 filters features using lightweight depthwise convolutions [28]. The VGG-19 is a deep learning convolutional neural network (CNN) architecture for image classification, with 16 convolutional layers and 3 fully connected layers [29]. Common convolutional neural network models, such as MobileNet v2 and VGG-19, are frequently utilized as pre-training network models for various transfer learning tasks. Consequently, these two network pre-training models are employed to implement image classification tasks in the experiment with the dataset of the custom six persons' dataset captured by a home camera, as shown in Figure 6.

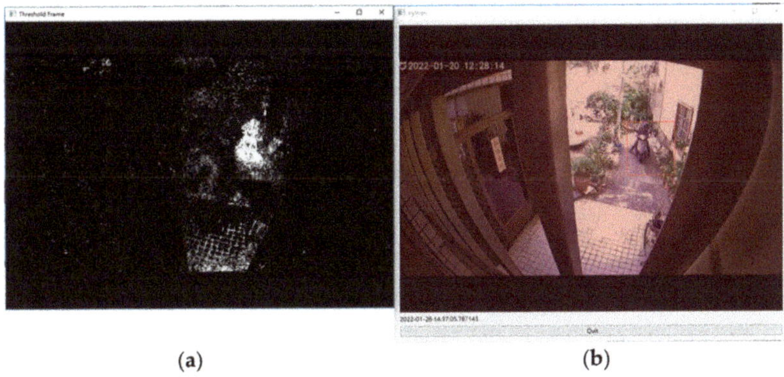

(a)   (b)

Figure 6. The custom 6 persons' dataset captured by a home camera: (**a**) frame difference result and (**b**) motion detection result.

Figure 7 illustrates the structure and parameters of the neural network test models based on MobileNet v2 and VGG-19 models. The pre-trained model is reused by simply removing the final layer and using it to classify new image categories. Layers are added to a model that has already been trained and they are frozen to prevent losing the information they contain during future training processes. On top of the frozen layers, trainable layers are added. In the test model based on the MobileNet v2 pre-trained model, the total params are 2,626,854; the trainable params are 368,870 and the nontrainable params are 2,257,984. In the test model based on the VGG-19 pre-trained model, the total params are 20,172,070; the trainable params are 147,686 and the nontrainable params are 20,024,384. The trained neural network model is saved with the TensorFlow model file format, which includes variable index (variables.index), variable value (variables.data-00000-of-00001), and GraphDef (*.pb) files. The GraphDef (*.pb) format contains serialized data and calculation graphs of protobuf objects. The operating system of the computer equipment used by the client is ubuntu, and the file difference comparison (*diff*) and neural network model patch file merge test are performed. In the experiment, the difference comparison tool used is hdiffz, and the correction tool used in the neural network model patch file is hpatchz [30]. The experiment explores the original model size of the neural network model, the size of the neural network model patch file after the difference comparison, and the data saving ratio.

```
Model: "sequential"
```

Layer (type)	Output Shape	Param #
vgg19 (Functional)	(None, 7, 7, 512)	20,024,384
conv2d (Conv2D)	(None, 5, 5, 32)	147,488
dropout (Dropout)	(None, 5, 5, 32)	0
global_average_pooling2d (Gl	(None, 32)	0
dense (Dense)	(None, 6)	198

```
Total params: 20,172,070
Trainable params: 147,686
Non-trainable params: 20,024,384
```

(**a**)

```
Model: "sequential_2"
```

Layer (type)	Output Shape	Param #
mobilenetv2_1.00_224 (Functi	(None, 7, 7, 1280)	2,257,984
conv2d_2 (Conv2D)	(None, 5, 5, 32)	368,672
dropout_2 (Dropout)	(None, 5, 5, 32)	0
global_average_pooling2d_2 (	(None, 32)	0
dense_2 (Dense)	(None, 6)	198

```
Total params: 2,626,854
Trainable params: 368,870
Non-trainable params: 2,257,984
```

(**b**)

Figure 7. The structure and parameters of the neural network test models based on MobileNet v2 and VGG-19 models. (**a**) The test model based on MobileNet v2 model. (**b**) The test model based on VGG-19 model.

The experiment needs to upgrade the related files of the neural network model, variable index (variables.index), variable value (variables.data-00000-of-00001), and GraphDef (*.pb) files. After the difference comparison of model files (*diff*), the model publish server generates neural network model-related patch files. Then, the neural network model-related patch files are packaged and transmitted to edge servers and edge IoT devices. After the new neural network model is updated on the edge IoT devices, the new neural network model can be loaded for inference and prediction. The experimental results are shown in Table 1.

Table 1 shows the relevant file sizes of the MobileNet v2 and VGG-19 neural network models after transfer learning with the six categories of person images. Among them, the neural network-related files after diff comparison are based on the MobileNet v2 neural network. With the transfer learning techniques, the models share the same weights of the basic neural network model, which can reduce the transmission data after differential comparison. The VGG-19 neural network model has up to 19 layers, and its basic neural network model variable value reaches 80,109,019 bytes. After the differential comparison, only 1,778,387 bytes need to be transmitted, which greatly reduces the need to transmit the total amount of model data. The new neural network model with MobileNet v2 transfer learning can save up to 72.93% of data, while the new neural network model with VGG-19 transfer learning can save data as much as 97.76%.

Table 1. The experimental results of image classification with diff and patch.

Image Classification (6 Persons)	
Number of images in dataset	289
Number of dataset categories	6
Training set/test set ratio	8:2
Basic Neural Network Model-mobilenetv2	
Basic neural network model pb model size	3,217,201
Basic neural network model variable index size	16,627
Basic neural network model variable value size	9,125,379
Model name-mobilenetv2_6persons (based on mobilenetv2)	
Model PB model size after transfer learning	4,350,337
Model variable index size after transfer learning	15,979
Model variable value size after transfer learning	15,979
Neural network pb model size after difference comparison	380,385
Neural network variable index size after difference comparison	4306
Neural network variable value size after difference comparison	4,464,323
The ratio of data saved after difference comparison	0.7293753813
The diff processing time	1.376 s
The patch processing time	0.004 s
Basic Neural Network Model-vgg19_base	
Basic neural network model pb model size	353,484
Basic neural network model variable index size	2285
Basic neural network model variable value	80,109,019
Model name-vgg19_6persons (based on vgg19_base)	
Model PB model size after transfer learning	537,634
Model variable index size after transfer learning	3434
Model variable value size after transfer learning	81,885,560
Neural network pb model size after difference comparison	64,741
Neural network variable index size after difference comparison	1243
Neural network variable value size after difference comparison	1,778,387
The ratio of data saved after difference comparison	0.9776240877
The diff processing time	8.439 s
The patch processing time	0.013 s

The generated neural network model-related files, i.e., variable index (variables.index), variable value (variables.data-00000-of-00001), and GraphDef (*.pb) files, are all compared with the original file with the SHA256 checksum, which verifies whether it is consistent or not after the patch update. The experimental verification monitors the checksum codes of the neural network model-related files, which are all correct, indicating that all the neural network model-related files are successfully deployed and updated. The time required for comparing (diff) the different file part among the new model with the six categories of person images and the original MobileNet v2 model is about 1.376 s and the patch time required for recovering (patch) the new model with the MobileNet v2 patch file is about 0.004 s in the experiment machine. The time required for comparing (diff) the difference among the new model with the six categories of person images and the original VGG-19 model is about 8.439 s and the patch time required for recovering (patch) the new model with the VGG-19 patch file is about 0.013 s in the experiment machine.

For investigating the system performance, the edge IoT node number is changed from 100 to 1000 for evaluating the simulation performance. In the simulation environment setup, the number of edge servers is set to 4; the available download bandwidth from a cloud server to edge server is set to 100 Mbps; the available download bandwidth from an edge server to an edge IoT is also set to 100 Mbps. The cloud to edge server's propagation delay is set to 300 ms and the edge server to edge IoT node's propagation delay is set to 30 ms. Figure 8 shows the simulation results. Label 'vgg_cloud' means that for a cloud server to transmit the VGG-19 model to 1000 edge IoT devices, it needs about 13,237 s. For the proposed diff model publish system, it only needs about 225 s to deploy new VGG-19 models to 1,000 edge IoT devices. For MobileNet v2, a similar result can be observed. The simulation results show that the proposed collaborative edge computing platform can speed up the deployment of artificial intelligence services.

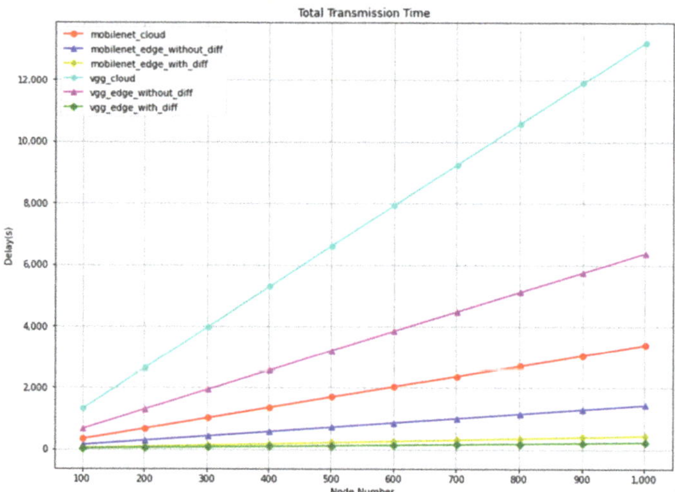

Figure 8. The simulation results of the proposed collaborative edge computing platform.

5. Discussion and Conclusions

Edge AI makes it possible to execute AI inference tasks locally without connecting to the cloud. Machine learning models must be quickly deployed on edge devices. Many cloud service providers provide AI edge computing deployment platforms for IoT devices, including Microsoft Azure IoT Edge [6], AWS Greengrass [3], Google's Cloud IoT Edge [4], and IBM Watson IoT Platform Edge [5].

When a neural network model is trained in the Microsoft Azure IoT Edge architecture, the complete neural model file must be repackaged into a new docker image before being deployed to IoT devices, which takes time and requires a sizable amount of data [6]. Machine learning models that are created, trained, and optimized in the cloud and run inference locally on devices are part of AWS IoT Greengrass [3]. Data gathered from the IoT devices can be sent back to AWS SageMaker where they can be used to continuously improve the quality of machine learning models [31]. AWS IoT Greengrass is capable of deploying components to IoT devices. Each IoT device operates on a combination of the software from the deployments that target the devices. Deployments to the same target device, however, overwrite any previous deployment components there. When a deployment for a target device is updated, the outdated components are swapped out for the updated components [32].

Google IoT Core contains two modules: (a) Device Manager allows users to set-up, authenticate, configure, and control IoT devices remotely; (b) Protocol Bridge operating with MQTT and HTTP protocols is in charge of service connectivity. Cloud Pub/Sub data are redirected to Google cloud services. For edge computing and AI in the Google IoT ecosystem, Google's Cloud IoT Edge is performed via its branded Edge TPU chip [33]. With the Coral platform for ML at the edge, Google's Cloud TPU and Cloud IoT are enhanced to offer an end-to-end (cloud-to-edge, hardware + software) infrastructure that makes it easier for clients to develop AI-based solutions. The Coral platform offers a full developer toolkit in addition to its open-source TensorFlow Lite programming environment, allowing users to create models or retrain a number of Google AI models for the Edge TPU, combining

Google's experience in both AI and hardware. However, the deployment method is not extensively defined [34].

IBM Edge Application Manager (IEAM) is designed specifically for edge node management to minimize deployment risks and to manage the service software lifecycle on edge nodes fully autonomously. Software developers develop and publish edge services to the management hub. Administrators define the deployment policies that control where edge services are deployed. IEAM publishes an existing Docker image as an edge service, creates an associated deployment pattern, and registers IoT edge nodes to run that deployment pattern. Similar to Microsoft Azure IoT Edge [6], the complete neural model file must be packaged into a new docker image before being deployed to IoT devices, which takes time and requires a sizable amount of data [35].

Existing edge IoT deployment platforms do not take into account the issue of sharing pre-trained models when deploying trained models, which leads to significant data consumption. Our work is different from the above system, and the proposed Cloud-Edge Smart IoT architecture and model publish system aims to solve the problems of deploying large numbers of neural network models to IoT devices and reducing the computing resource requirements of cloud servers. Only a small amount of data are required to be transmitted to all edge IoT devices for new model updating, and the entire neural network model can be updated. Existing edge computing architectures do not support updating the weights of various neural network models, nor are they optimized for storage, updating, and transmission of various neural network models for a large number of edge IoT devices. The proposed collaborative edge computing platform uses pre-trained neural layers and weight information to train and update the neural network model. Through differential comparison, the edge server only needs to send a small-part neural network model patch file to complete the deployment of new neural network models on IoT devices. The experimental results show that the proposed neural network model weight comparison mechanism can speed up the deployment of artificial intelligence services.

Author Contributions: Conceptualization, T.-H.H.; software, T.-H.H.; data curation, T.-H.H.; methodology, T.-H.H. and Z.-H.W.; validation, T.-H.H., Z.-H.W. and A.R.S.; writing—original draft, T.-H.H.; writing—review and editing, A.R.S. All authors have read and agreed to the published version of the manuscript.

Funding: This research was funded by the Ministry of Science and Technology (MOST), Taiwan, Grant Nos. MOST 110-2221-E-218-018 and MOST 109-2221-E-218-019.

Acknowledgments: The authors wish to express their gratitude to the Ministry of Science and Technology (MOST) for support this research.

Conflicts of Interest: The authors declare no conflict of interest.

References

1. Ai, Y.; Peng, M.; Zhang, K. Edge computing technologies for Internet of Things: A primer. *Digit. Commun. Netw.* **2018**, *4*, 77–86. [CrossRef]
2. IoT Edge | Cloud Intelligence | Microsoft Azure. Available online: https://azure.microsoft.com/en-us/services/iot-edge/ (accessed on 22 June 2022).
3. AWS IoT Greengrass Documentation. Available online: https://docs.aws.amazon.com/greengrass/index.html (accessed on 22 June 2022).
4. Google Cloud IoT-Fully Managed IoT Services. Available online: https://cloud.google.com/solutions/iot (accessed on 22 June 2022).
5. IBM Watson IoT Platform. Available online: https://internetofthings.ibmcloud.com/internetofthings.ibmcloud.com (accessed on 22 June 2022).
6. Zhang, Y. Deploy Machine Learning Models on Azure IoT Edge. Available online: https://github.com/microsoft/deploy-MLmodels-on-iotedge/commits?author=YanZhangADS (accessed on 22 June 2022).
7. Weiss, K.; Khoshgoftaar, T.M.; Wang, D. A Survey of Transfer Learning. *J. Big Data* **2016**, *3*, 9. [CrossRef]
8. Leroux, S.; Bohez, S.; Verbelen, T.; Vankeirsbilck, B.; Simoens, P.; Dhoedt, B. Transfer Learning with Binary Neural Networks. *arXiv* **2017**, arXiv:1711.10761.
9. Pan, S.J.; Yang, Q. A Survey on Transfer Learning. *IEEE Trans. Knowl. Data Eng.* **2010**, *22*, 1345–1359. [CrossRef]
10. Transfer Learning for Machine Learning. Available online: https://www.seldon.io/transfer-learning (accessed on 21 June 2022).

11. Transfer Learning Guide: A Practical Tutorial with Examples for Images and Text in Keras. Available online: https://neptune.ai/blog/transfer-learning-guide-examples-for-images-and-text-in-keras (accessed on 21 June 2022).
12. Wang, C.; Mahadevan, S. Heterogeneous Domain Adaptation Using Manifold Alignment. In Proceedings of the Twenty-Second International Joint Conference on Artificial Intelligence, Barcelona, Spain, 16 July 2011; AAAI Press: Palo Alto, CA, USA, 2011; Volume 2, pp. 1541–1546.
13. Li, W.; Duan, L.; Xu, D.; Tsang, I.W. Learning with Augmented Features for Supervised and Semi-Supervised Heterogeneous Domain Adaptation. *IEEE Trans. Pattern Anal. Mach. Intell.* **2014**, *36*, 1134–1148. [CrossRef] [PubMed]
14. Zhu, Y.; Chen, Y.; Lu, Z.; Pan, S.J.; Xue, G.-R.; Yu, Y.; Yang, Q. Heterogeneous Transfer Learning for Image Classification. In Proceedings of the Twenty-Fifth AAAI Conference on Artificial Intelligence, San Francisco, CA, USA, 7 August 2011; AAAI Press: Palo Alto, CA, USA, 2011; pp. 1304–1309.
15. Shi, S.; Wang, Q.; Xu, P.; Chu, X. Benchmarking State-of-The-Art Deep Learning Software Tools. In Proceedings of the 2016 7th International Conference on Cloud Computing and Big Data (CCBD), Macau, China, 16–18 November 2016.
16. Zhang, Q.; Yang, L.T.; Chen, Z.; Li, P.; Deen, M.J. Privacy-Preserving Double-Projection Deep Computation Model with Crowdsourcing on Cloud For Big Data Feature Learning. *IEEE Internet Things J.* **2017**, *5*, 2896–2903. [CrossRef]
17. Garcia Lopez, P.; Montresor, A.; Epema, D.; Datta, A.; Higashino, T.; Iamnitchi, A.; Barcellos, M.; Felber, P.; Riviere, E. Edge-centric Computing: Vision and Challenges. *ACM SIGCOMM Comput. Commun. Rev.* **2015**, *45*, 37–42. [CrossRef]
18. Dolui, K.; Datta, S.K. Comparison of Edge Computing Implementations: Fog Computing, Cloudlet and Mobile Edge Computing. In Proceedings of the 2017 Global Internet of Things Summit (GIoTS), Geneva, Switzerland, 6–9 June 2017.
19. Bonomi, F.; Milito, R.; Zhu, J.; Addepalli, S. Fog Computing and its Role in the Internet of Things. In Proceedings of the MCC Workshop on Mobile Cloud Computing, Helsinki, Finland, 17 August 2012.
20. Beck, M.T.; Werner, M.; Feld, S.; Schimper, S. Mobile Edge Computing: A Taxonomy. In Proceedings of the 6th International Conference on Advances in Future Internet, Lisbon, Portugal, 16–20 November 2014.
21. Satyanarayanan, M.; Bahl, P.; Caceres, R.; Davies, N. The Case for Vm-Based Cloudlets in Mobile Computing. *IEEE Pervasive Comput.* **2009**, *8*, 14–23. [CrossRef]
22. Sarkar, D. (DJ) A Comprehensive Hands-On Guide to Transfer Learning with Real-World Applications in Deep Learning. Available online: https://towardsdatascience.com/a-comprehensive-hands-on-guide-to-transfer-learning-with-real-world-applications-in-deep-learning-212bf3b2f27a (accessed on 22 June 2022).
23. Django. Available online: https://djangoproject.com (accessed on 5 December 2020).
24. Celery-Distributed Task Queue. Available online: https://docs.celeryproject.org/en/stable/#celery-distributed-task-queue (accessed on 8 November 2020).
25. Redis. Available online: https://redis.io/ (accessed on 9 March 2021).
26. MQTT Version 3.1.1. Available online: http://docs.oasis-open.org/mqtt/mqtt/v3.1.1/os/mqtt-v3.1.1-os.html (accessed on 4 July 2022).
27. Eclipse Mosquitto. Available online: https://mosquitto.org/ (accessed on 10 March 2021).
28. Sandler, M.; Howard, A.; Zhu, M.; Zhmoginov, A.; Chen, L.-C. MobileNetV2: Inverted Residuals and Linear Bottlenecks. In Proceedings of the 2018 IEEE/CVF Conference on Computer Vision and Pattern Recognition, Salt Lake City, UT, USA, 18–22 June 2018; pp. 4510–4520.
29. Simonyan, K.; Zisserman, A. Very Deep Convolutional Networks for Large-Scale Image Recognition. In Proceedings of the 3rd International Conference on Learning Representations, ICLR 2015, San Diego, CA, USA, 7–9 May 2015.
30. HDiffPatch. Available online: https://github.com/sisong/HDiffPatch (accessed on 10 May 2021).
31. AWS Greengrass Machine Learning Inference-Amazon Web Services. Available online: https://aws.amazon.com/greengrass/ml/ (accessed on 25 June 2022).
32. Deploy AWS IoT Greengrass Components to Devices-AWS IoT Greengrass. Available online: https://docs.aws.amazon.com/greengrass/v2/developerguide/manage-deployments.html (accessed on 25 June 2022).
33. Making Sense of IoT Platforms: AWS vs. Azure vs. Google vs. IBM vs. Cisco. AltexSoft. Available online: https://www.altexsoft.com/blog/iot-platforms/ (accessed on 25 June 2022).
34. Edge TPU-Run Inference at the Edge. Available online: https://cloud.google.com/edge-tpu (accessed on 25 June 2022).
35. Transform Image to Edge Service. Available online: https://prod.ibmdocs-production-dal-6099123ce774e592a519d7c33db8265e-0000.us-south.containers.appdomain.cloud/docs/en/eam/4.3?topic=SSFKVV_4.3/OH/docs/developing/transform_image.html (accessed on 25 June 2022).

 *electronics*

Article

Application of Generative Adversarial Network and Diverse Feature Extraction Methods to Enhance Classification Accuracy of Tool-Wear Status

Bo-Xiang Chen, Yi-Chung Chen *, Chee-Hoe Loh, Ying-Chun Chou, Fu-Cheng Wang and Chwen-Tzeng Su

Department of Industrial Engineering and Management, National Yunlin University of Science and Technology, No. 123 University Road, Section 3, Douliou, Yunlin 64002, Taiwan; m10821012@yuntech.edu.tw (B.-X.C.); d10721004@yuntech.edu.tw (C.-H.L.); d10921002@yuntech.edu.tw (Y.-C.C.); fcwang@yuntech.edu.tw (F.-C.W.); suct@yuntech.edu.tw (C.-T.S.)
* Correspondence: chenyich@yuntech.edu.tw or mitsukoshi901@gmail.com

Abstract: The means of accurately determining tool-wear status has long been important to manufacturers. Tool-wear status classification enables factories to avoid the unnecessary costs incurred by replacing tools too early and to prevent product damage caused by overly worn tools. While researchers have examined this topic for over a decade, most existing studies have focused on model development but have neglected two fundamental issues in machine learning: data imbalance and feature extraction. In view of this, we propose two improvements: (1) using a generative adversarial network to generate realistic computer numerical control machine vibration data to overcome data imbalance and (2) extracting features in the time domain, the frequency domain, and the time–frequency domain simultaneously for modeling and integrating these in an ensemble model. The experiment results demonstrate how both proposed modifications are reasonable and valid.

Keywords: tool wear; data imbalance; GAN; ensemble learning

Citation: Chen, B.-X.; Chen, Y.-C.; Loh, C.-H.; Chou, Y.-C.; Wang, F.-C.; Su, C.-T. Application of Generative Adversarial Network and Diverse Feature Extraction Methods to Enhance Classification Accuracy of Tool-Wear Status. *Electronics* 2022, 11, 2364. https://doi.org/10.3390/electronics11152364

Academic Editor: Martin Reisslein

Received: 1 June 2022
Accepted: 26 July 2022
Published: 28 July 2022

Publisher's Note: MDPI stays neutral with regard to jurisdictional claims in published maps and institutional affiliations.

1. Introduction

Tool management for computer numerical control (CNC) has long been a topic of focus for manufacturers. Tools are worn down as they are used. Below a certain degree of wear, they can still function normally. However, once the wear reaches the threshold, it will no longer function normally and may even damage the products. Manufacturers must therefore carefully monitor tool wear in CNC machinery and replace the tools when the extent of wear approaches the threshold. In the past, the timing at which tools should be replaced was difficult to determine. Manufacturers mainly had to rely on the experience of onsite personnel, who determined the timing based on the sound of cutting or the statuses of the previously processed product. This approach is inconvenient: an experienced worker must be monitoring the machinery at all times during operation, and even then, tools may be replaced too early or too late. The former means discarding tools when they can still be used, which is a waste of resources. The latter may result in damaged products, which reduces the yield and incurs additional costs. To avoid these issues, researchers have developed the Prognostics and Health Management guidelines [1–3] to assist factories in predicting and managing the health status of machines. This framework comprises the six following steps: data processing, feature extraction, diagnostics, prognostics, decision support, and feedback and learning. Among these, diagnostics (identification of tool wear state) and prognostics (prediction of remaining tool life) are the most frequently discussed. These two steps are key to the success of analysis, as it is impossible to manage the status of machines if these two steps are not executed well. Both diagnostics and prognostics are significantly influenced by variables such as the machine type, tool type, and environment. In this study, we focus on diagnostics, which we define as directly determining the wear status of a tool (i.e., rapid initial wear, uniform wear, or failure wear) based on vibration or sound data (see Figure 1).

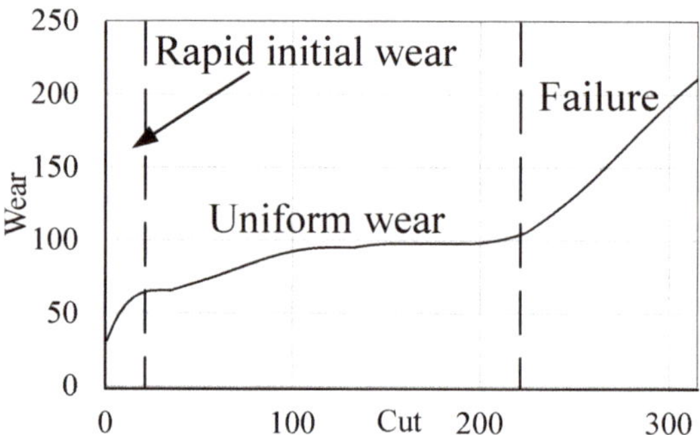

Figure 1. Examples of three types of tool-wear statuses.

Most existing methods for tool wear analysis employ various sensors such as vibration sensors, acoustic emission sensors, and torque sensors to collect data from machine operations. Then, relevant information called features is extracted from the collected data. The features are input into various machine learning models for modeling [4,5]. For example, Chen et al. [6] used a logistic regression-based model to analyze the vibration signals for monitoring the statuses of tools. Kong et al. [7] proposed a Gaussian process regression model to predict the wear of the tool. Benkedjouh et al. [8] used multiple sensors to collect data and used these data to understand the health of cutting tool. Cai et al. [9] proposed a proportional covariate model for analyzing the vibration signals and thus monitored the reliability of the tools. Zhu and Liu [10] and Yu et al. [11] applied the Markov-model-based method to monitor the tool wear and predict the statuses of tools, respectively. Later, some deep learning models were developed to assess the statuses of tools during the manufacturing process. For example, Kurek et al. [12] employed a convolutional neural network to analyze the drill wear. Rohan et al. [13] used convolutional neural networks to detect and diagnose the faults of an industrial robot. To predict tool wear, Zhang et al. [14] used a long short-term memory model, whereas Cao et al. [15] combined derived wavelet frames with a convolutional neural network. In contrast, Sun et al. [5] utilized an auto-encoder method. Chen et al. [16] designed a framework based on the radial basis function and deep recurrent neural networks to swiftly generate lightweight models for the prediction of tool lifespan.

The researchers above all claimed that their approaches could successfully analyze tool-wear status and lifespan. However, we found that most of these studies focused on model development and neglected two fundamental issues in machine learning: data imbalance and feature extraction. The quantity of data for the three tool-wear statuses (i.e., rapid initial wear, uniform wear, and failure wear) will always be imbalanced. For example, in Figure 1, the quantities of the tool-wear statuses in descending order are uniform wear > failure wear > rapid initial wear. In practice, data on failure wear are even less, as tools are replaced as soon as failure wear is detected. Therefore, most manufacturers will only have access to the datasets featuring data imbalance. Few studies have addressed this topic. Carino et al. [17] proposed using incremental learning to debug data imbalance for the friction test. Brito et al. [18] suggested using unsupervised artificial intelligence techniques to improve the identification of imbalanced datasets while Miao et al. [19] proposed using deep supervision to introduce a surrogate loss function based on the Matthews correlation coefficient. Similarly, Rohan [13] and Rohan et al. [20] proposed using a generative adversarial network model with robotic arms. However, the methodology proposed by [13,17,20] is not specifically designed for tool wear, and while the methodology proposed by [18,19] is applicable to tool wear, its solution is limited to a specific model, which is not generalizable. The current paper therefore proposes a novel approach to data imbalance that is widely applicable. With regard to feature extraction, most algorithms only implement a single feature extraction method, such as one that only considers the features of tool-wear data in the time domain, the frequency domain, or the time–frequency domain. However, whether a single feature extraction method can completely extract features crucial to tool-wear status remains undetermined. This is not only because the features of wear data generated by different CNC machines may differ, but also because the features of wear data generated by the same CNC machine may also vary significantly with the workpiece. The features of wear data from different wear statuses may also vary. For some wear statuses, time features may be better classified in the time domain; for other wear statuses, the frequency or time–frequency domain may be required. Thus, we propose that multiple feature extraction methods are imperative in modeling tool wear.

This study proposes two approaches to overcome the aforementioned issues: (1) using a generative adversarial network (GAN) to generate data to overcome data imbalance and (2) extracting three types of features in the target algorithm and integrating these in an ensemble model. With regard to the first approach, GANs are a type of deep learning model in the field of artificial intelligence. They use two deep learning networks, namely a generator and a discriminator, that learn from each other to generate realistic data. The generator receives input comprising a set of vectors and then creates data that have features similar to those of historical data, whereas the discriminator determines whether the data generated by the generator are realistic. During the training process, the generator and the discriminator are continuously trained to outperform the other; the better the generator is at creating realistic data, the better the discriminator must become at identifying fake data. After a series of rounds, the features of the data produced by the generator become increasingly similar to those of historical data, thereby achieving the objective of the GAN. The GAN has been widely applied to various topics involving data generation. For instance, Goodfellow [21] designed a basic GAN framework to generate real-looking handwritten numbers and human faces. Karras et al. [22] proposed a novel progressive GAN and verified that it can greatly improve human face generation. Yadav et al. [23] developed a cyclic synthesized attention-guided GAN to further optimize virtual human face generation. Shi et al. [24] designed a GAN that transforms 2D human face images into 3D images. Fang et al. [25] developed a GAN to produce human faces from human speech fragments. Chen et al. [26] proposed a GAN that uses two discriminators at the same time to repair images. Wei et al. [27] presented an occlusion-aware warping GAN to overcome the issue of blocked human images in videos. In addition to human face recognition, a number of recent studies have applied GAN to generate manufacturing data. For example, Tagawa et al. [28] used a GAN to reconstruct sound signals and detect abnormalities in noisy factory environments. Zhang et al. [29] proposed a multi-view GAN to generate images of real vehicles from skeleton views. Gan et al. [30] employed a GAN to enhance the detection rates of an automatic leather patch defection system. Gu et al. [31] utilized a conditional GAN to generate samples of rolling bearing failures and thereby enhance the accuracy of detecting failure based on vibration signals from rolling bearings. In the current paper, we applied a GAN to generate realistic CNC machine operating data (including vibration signals and sound signals) to resolve data imbalance.

In terms of feature extraction, we surveyed relevant studies to identify the three domains most commonly employed for modeling: the time, frequency, and time–frequency domains. We established a deep learning model for each type of extracted feature and then employed an ensemble learning model to integrate the results. We postulated that extracting various types of features will offer an advantage in tool-wear status classification by processing wear problems comprehensively.

In our framework, we first cleaned the collected data and used a GAN to produce additional operating data. Then, we obtained features from the time, frequency, and time–frequency domains. Third, we used three deep learning models to model the three types of features; the output of each model was the classification of the current tool-wear status as rapid initial wear, uniform wear, or failure wear. Finally, we used an ensemble learning model to integrate the results of the three models to output the current tool-wear status to the user. We conducted several experiments to verify the performance of the proposed approaches.

The remainder of this paper is structured as follows. Section 2 introduces relevant literature on tool wear and Section 3 outlines the framework of the proposed approaches. Section 4 presents our experiment simulations and Section 5 contains the conclusion and directions for future work.

2. Related Work

This chapter reviews three important topics: (1) tool-wear statuses; (2) data fields suitable for tool wear predictions; and (3) existing tool wear prediction methods.

2.1. Tool-Wear Statuses

Tool wear can be divided into three phases [32]: rapid initial wear, uniform wear, and failure. These are depicted in Figure 1. In the first phase, tool blades may not be of uniform lengths and blade edges are very sharp. Thus, wear in this phase is rapid. In the second phase, differences in tool length have evened out, creating a larger area to bear force, which reduces the pressure. Thus, wear in this phase is slower and steadier, with no sharp fluctuations. In the final phase, the tool has become blunt, thereby increasing the cutting resistance, required cutting power, and cutting temperature. The wear rate therefore significantly increases, and the probability of failure is high. In practice, tools are replaced before they reach this phase to protect process quality. To achieve consistency and comparability in tool-wear judgment, ISO 8688-2 suggests that tools should be replaced if the average wear of multiple tools exceeds 0.3 mm or if the wear of a single tool exceeds 0.5 mm [33].

2.2. Data Fields Suitable for Tool Wear Predictions

Researchers have demonstrated which data fields can be used to effectively predict tool-wear statuses; these include sound, vibrations, and electric currents. For instance, Erturk et al. [34] indicated that process parameters such as cutting speed, cutting time, and cutting depth all exert influence on tool wear. Bhuiyan et al. [35] utilized an acoustic emission sensor to collect the soundwave signals produced by a cutting tool to analyze tool-wear statuses. Dolinsek et al. [36] similarly used an acoustic emission sensor to examine the relationship between tool-wear statuses and workpiece material. Bhuiyan et al. [37] speculated that when tools become dull, the rotational speed slows down, causing the machine to increase its electrical current to reach the required speed. They therefore used the relationship between the rotational speed and electrical current to predict tool wear. In recent years, a number of studies have used vibration signals to predict tool wear [6,9]. Despite good results, some researchers insist that a single sensor fails to provide a comprehensive evaluation of tool wear. For example, Benkedjouh et al. [8] used acoustic emission sensors, accelerometers, and force sensors to collect data on tool wear.

2.3. Existing Tool Wear Prediction Methods

Approaches to tool wear prediction research can be divided into two categories: early machine-learning methods and recent deep-learning methods. The former approach has been widely applied in a range of contexts. Li et al. [38] used a random forest and a multiple linear regression model to analyze vibration signals. Kong et al. [7] proposed a novel Gaussian regression model. Cai et al. [9] proposed a proportional covariate model for vibration signals. Gomes et al. [39] employed a support vector machine to analyze the vibration data from milling manufacturers. Mohanraj et al. [40] also used a support vector machine for milling data but included a decision tree for feature selection to increase the monitoring accuracy. Jalali et al. [41] used a support vector machine to monitor the ball bearing failure with a genetic algorithm for feature selection. Markov models and artificial neural networks are also popular. For instance, Zhu and Liu [10] and Yu et al. [11] utilized a Markov model to predict the tool statuses, while Corne et al. [42] and Hesser et al. [43] used artificial neural networks to, respectively, monitor drilling processes and tool wear.

The emergence of DLMs brought increased accuracy to tool wear predictions. Zhang et al. [14] and Zhao et al. [44] used LSTM models to monitor machine health. Kurek et al. [12] analyzed the drill head wear using a convolutional neural network. Cao et al. [15] and Cheng et al. [45] first applied the wavelet transform to sound or vibration signals and then input the results into a convolutional neural network. Other examples include Sun et al. [5], who used an auto-encoder method, and Zhao et al. [46], who employed a gated recurrent unit-based approach to perform the gear and shaft malfunction detection. These studies demonstrate the superiority of DLMs over conventional machine learning.

3. Frameworks

The framework in this paper is divided into two stages, as shown in Figure 2. In the first stage, collected tool vibration data are cleaned using linear interpolation to fill in the missing values and the data are organized into a temporal matrix format to serve as GAN input. This represents Step 1. In Step 2, the GAN model is established and trained to solve the common data imbalance issue in tool-wear problems. The second stage includes Step 3, in which three different methods are employed to obtain the features of tool-wear status classifications and solve the problem of the unsuitability of single-classification feature selection methods for all tool-wear problems. In Step 4, a convolutional neural network (CNN) is established and trained for each feature selection method to classify the tool-wear status. The final step of the second stage is Step 5, in which a shallow neural network (SNN) is used to perform ensemble learning with the outputs of the three CNNs established in the previous step. Below, we describe all five steps in detail.

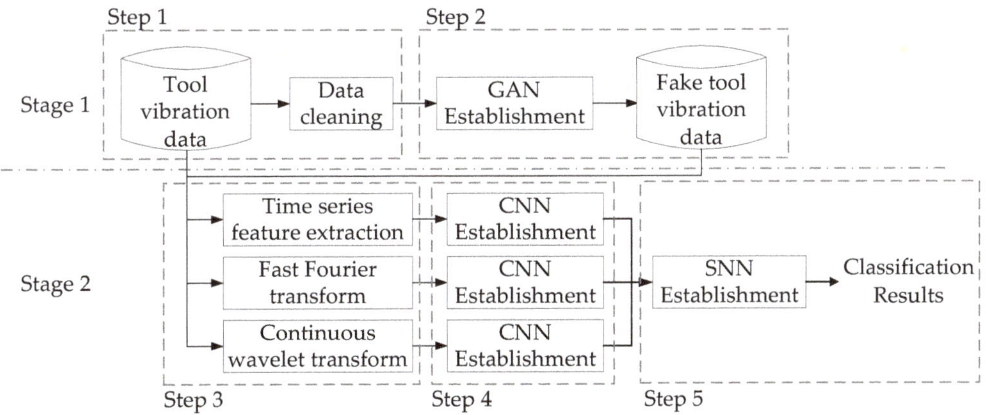

Figure 2. Flow chart of research framework.

3.1. Introduction to the Dataset and the Methods for Data Cleaning

The dataset used in this study was from the 2010 PHM Data Challenge [16,47]. It contains data collected during cutting using a CNC machine, with a sensor data collection rate of 50 k/Hz, 6 mm ball nose tungsten carbide cutters, and HRC52 stainless steel workpieces. The parameters of the machine-cutting experiment were a spindle speed of 10,400 RPM, a feed rate of 1555 mm/min, and a cutting depth of 0.2 mm. Data were included from six cutters. The organizers of the challenge selected three of the six cutters as training sets and provided the wear value after each cut as well as the vibration data from the cutting processes. The data from the three remaining cutters served as test sets. For these cutters, the wear values after each cut were not provided, so we only employed the three datasets serving as training sets in this study.

Each cutter was used to make 315 cuts, and for each cut, various features were recorded, as shown in Table 1. The seven columns in the table present acceleration in the x axis, acceleration in the y axis, acceleration in the z axis, vibration in the x axis, vibration in the y axis, vibration in the x axis, and acoustic emission. As the cutting time varied with each cut, the quantity of data collected ranged from 100,000 to 300,000 items. As for the tool-wear status, we divided the wear values into rapid initial wear (0~$\leq$66); uniform wear (>66~$\leq$165); and failure wear (>165), as suggested by experts. As shown in Figure 3, the amount of uniform wear data was far greater than the amounts of rapid initial wear and failure wear, which makes the dataset suitable for verifying the proposed algorithm.

Table 1. Dataset from 2010 PHM data challenge.

	Acceleration in x Axis	Acceleration in y Axis	Acceleration in z Axis	x Vibrations	y Vibrations	z Vibrations	Acoustic Emission
1	0.704	−0.387	−1.084	0.018	0.031	0.027	−0.004
2	0.772	−0.573	−1.153	−0.056	−0.057	−0.058	−0.004
3	0.828	−0.673	−1.242	0.037	0.019	0.031	−0.004
...	...	...	...	...	...	...	...
127,399	0.207	0.483	0.292	0.111	0.114	0.125	−0.004

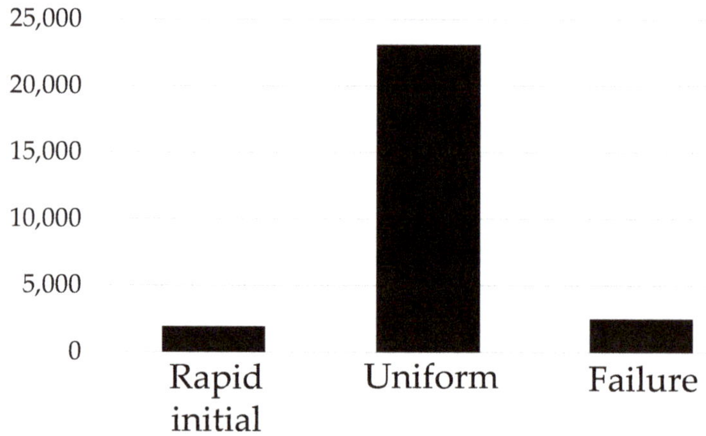

Figure 3. Distributions of tool-wear classifications.

Our data cleaning process includes two parts, namely the linear imputation process and the data conversion process. We now introduce these parts in the following.

Linear imputation: the vibration data used in this study were collected from sensors installed on CNC machines. However, voltage instability, network equipment failure, or issues with the sensors themselves during machine operation may have caused missing values in the collected data, and such data cannot be used for model training. As suggested in past studies [48], we employed linear imputation to fill in the missing values, as follows:

$$x_n = x_{n-1} + n\Delta, \tag{1}$$

$$\Delta = \frac{x_{t+1} - x_0}{t+1}, \tag{2}$$

where t denotes the range in need of imputation and n represents the nth item of data in need of imputation.

Data conversion: GAN training in past studies was achieved using images. We attempted to conduct training using time series. To input the time series data into the GAN for training, we converted the post-linear imputation data into a temporal matrix format.

3.2. Use of GAN to Generate Realistic Vibration Data to Overcome Data Imbalance

This section introduces the GAN framework and training method of the target approach. Figure 4 displays the framework of the target GAN, which includes a generator and discriminator. The generator receives random noise and then generates a set of tool vibration data, whereas the discriminator receives the generated vibration data and the original vibration data and determines whether the generated vibration data are similar to the true data. The resulting determination is then provided to the generator as feedback. Based on the feedback, the generator then generates even more realistic tool vibration data for the discriminator to assess. The entire training process is repeated until the discriminator cannot determine whether the generated data are real or not.

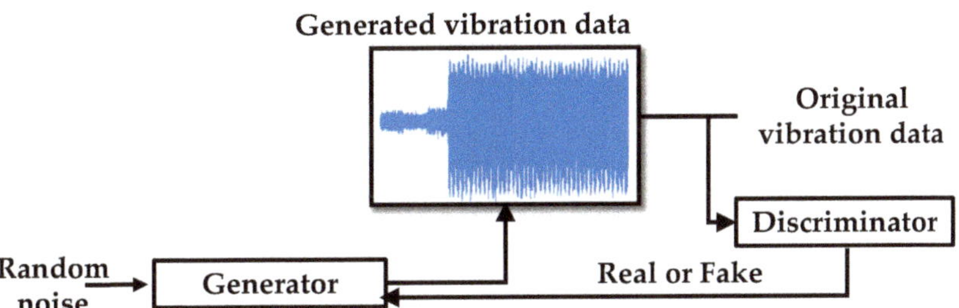

Figure 4. Proposed GAN framework.

The generator used in this study comprised three types of layers: an input layer, multiple upsampling layers, and multiple convolution layers. Below, we explain the input and the formulas of the neurons in each layer in the generator in detail. First, regarding the input data, suppose that the random data have n items of data and m features. Thus, the input is a $n \times m$ random number matrix, the elements of which are normally distributed random numbers between 0 and 1.

The input layer places the input data in the model. The input data of this layer are the output results, as follows:

$$O_i^{input} = x_i^{input}, \tag{3}$$

where O denotes the neuron output, x represents the input data, and i is the ith input of the model.

The upsampling layers employ nearest-neighbor interpolation, copying existing data to augment the feature map so that the model can better learn the features of the data during training. With input n and sample size *size*, the formula of the upsampling layers is as follows:

$$O_i = x_{\lfloor i/size \rfloor}, \; i \in [0, n \times size), \; i \in N \tag{4}$$

where O_i represents the output of the ith neuron and x_i denotes the ith input.

In convolution layers, each node has k kernels of size m. Thus, the formula of the convolutional layers is as follows:

$$O_i = act\left(\sum_{k=0}^{k=m-1} x_{i+m-k} w_k + b_i \right), \tag{5}$$

where O_i represents the output of the ith neuron, x_i denotes the ith original input, W is the filter, and $act(\bullet)$ is an activation function.

Table 2 shows the structure of the GAN generator as applied to the 2010 PHM dataset. The random numbers we input initially formed a 125×128 matrix. Then, aside from the input layer, the model itself also included four sets of upsampling and convolution layers. Note that the number of sets was decided via trial and error. Due to the high degree of detail that we desired from the produced signals, we set the kernel size and stride at 3 and 2, respectively. Finally, for the activation function, we used LeakyReLU in the first three convolutional layers to ensure that the features in the inputs could be fully displayed without falling into the dead zone. Due to the output signal values, we adopted tanh for the last convolutional layer to obtain better results. Finally, because the original 2010 PHM dataset contained seven dimensions, we set the number of dimensions in the generator output data to seven.

Table 2. The structure of the generator.

Layer	Type	Output Size	Kernel Size	Stride	Activation Function
Input	Input	125×128	-	-	-
U1	Upsampling	250×128	-	-	-
C2	Convolutional	125×64	3	2	LeakyReLU
U3	Upsampling	250×64	-	-	-
C4	Convolutional	125×64	3	2	LeakyReLU
U5	Upsampling	250×64	-	-	-
C6	Convolutional	125×64	3	2	LeakyReLU
U7	Upsampling	250×64	-	-	-
Output	Convolutional	250×7	-	-	tanh

The discriminator used in this study employs a one-dimensional CNN, the input of which is real tool vibration data and the tool vibration data generated by the generator and the output is the degree of similarity between the two. The architecture of this CNN includes four types of layers: an input layer, multiple convolution layers, multiple batch normalization layers, and an output layer. The convolution layer and the batch normalization layer are used alternately.

The mathematical formulas of the input layer, convolution layers, and output layer are identical to those in the generator. The batch normalization layers help mitigate gradient vanishing and accelerate neural convergence, as follows:

$$O_i = \frac{x_i - \overline{x}}{\sqrt{Var(x_i)}} + \beta, \tag{6}$$

where O and x denote the output and input, respectively, i represents the ith neuron, and γ and β represent the scale and shift.

Once the generator and discriminator have been established, we used a backpropagation algorithm to train the two networks. The aim is to minimize generator loss, which means that the data generated by the generator are as close as possible to the ground truth, making it difficult for the discriminator to distinguish real data from generated data. The formula for generator loss is as follows:

$$GeneratorLoss = \frac{1}{N} \sum_{i=1}^{N} \log\left(1 - D\left(G\left(\mathbf{R}^i\right)\right)\right), \tag{7}$$

where $\mathbf{R}$ is a vector of random numbers, D and G represent the discriminator and generator, respectively, and N denotes the training data. With regard to the discriminator, the aim is to minimize discriminator loss, which means that the discriminator has the ability to distinguish real data from generated data. The formula for discriminator loss is as follows:

$$Discriminator = -\frac{1}{N} \sum_{i=1}^{N} \log D\left(X^i\right) - \frac{1}{N} \sum_{i=1}^{N} \log\left(1 - D\left(\hat{X}^i\right)\right), \tag{8}$$

where X denotes the original data, $\hat{X}$ represents the generator, D is the discriminator, and N denotes the training data.

Table 3 presents the structure of the GAN discriminator as applied to the 2010 PHM dataset. The realistic data generated by the generator and the real data were both input to the model. After the input layer, we used a convolutional layer to perform dimension reduction. We then used three sets of convolutional layers and batch normalization layers to check the generated data. The number of sets was also decided via trial and error. As for the kernel size and stride parameters, we adopted settings similar to those of the generator, setting them as 3 and from 1 to 2, respectively. Finally, for the activation function, the objective of the discriminator was to inspect whether the data features are reasonable. We therefore used LeakyReLU for all convolutional layers to ensure that the features in the inputs could be fully displayed without falling in the dead zone. Finally, we used a fully connected layer to gauge whether the output was realistic.

Table 3. The structure of the discriminator.

Layer	Type	Output Size	Kernel Size	Stride	Activation Function
Input	Input	250×7	-	-	-
C1	Convolutional	125×128	3	2	LeakyReLU
C2	Convolutional	63×128	3	2	LeakyReLU
BN3	Batch normalization	63×128	-	-	-
C4	Convolutional	63×64	3	2	LeakyReLU
BN5	Batch normalization	63×64	-	-	-
C6	Convolutional	63×64	3	1	LeakyReLU
BN7	Batch normalization	63×64	-	-	-
Output	Fully connected	1	-	-	-

3.3. Feature Selection

This section introduces three methods that we used to extract the features of the vibration signals. The first method is time series feature extraction, in which changes in the amplitudes of tool cutting in the same time interval are analyzed. The second method is fast Fourier transform (FFT), in which the relationship between amplitude and frequency during tool cutting is observed and analyzed. The last method is continuous wavelet transform, in which changes in the amplitudes in time and frequency during tool cutting are observed and analyzed.

3.3.1. Time Series Feature Extraction

To extract the important features of the time series, researchers have used overlapping windows of a fixed size to segment time series into equal lengths and then extracted various feature statistics (including maximum, minimum, mean, sum, average absolute deviation, root mean square error, and standard deviation) from each segmented window [49]. In this study, we used this approach to extract the statistical features of the vibration data in all seven domains. Below, we introduce the formulas for each feature, assuming that the original dataset can be expressed as $\mathbf{X} = [x_1, x_2, x_3, \ldots, x_n]$ and data $\mathbf{X}(t)$ in the window at time point t can be written as $\mathbf{X}(t)=[x_t, x_{t+1}, \ldots, x_{t+w}]$, where w is the length of the window:

1. $F_{Max}(t)$ is the maximum value in $X(t)$;
2. $F_{Min}(t)$ is the minimum value in $X(t)$;
3. $F_{avg}(t)$ is the mean of all values in $X(t)$;
4. $F_{sum}(t)$ is the sum of all values in $X(t)$;
5. $F_{mad}(t)$ is the degree of dispersion among all values in $X(t)$:

$$MAD = \frac{1}{|t|} \sum_{i=1}^{|t|} |x_i - m|, \qquad (9)$$

6. $F_{RMSE}(t)$ is the root mean square error of all values in $X(t)$:

$$RMSE = \sqrt{\frac{1}{|t|} \sum_{i=1}^{|t|} x_i^2}, \qquad (10)$$

7. $F_{std}(t)$ is the standard deviation of all of the values in $X(t)$:

$$STD = \sqrt{\frac{\sum_{i=1}^{n}(x_i - \bar{x})^2}{n-1}}, \tag{11}$$

3.3.2. Fast Fourier Transform (FFT)

Tool vibration data are continuous, and many researchers have analyzed such data using FFT [50–53]. FFT is an accelerated form of the discrete Fourier transform which converts time-domain data into the frequency domain for the convenience of users. Its formula is as follows:

$$x\left(e^{j\hat{\omega}_k}\right) = \sum_{n=0}^{L-1} x[n]e^{j\hat{\omega}_k n}, \tag{12}$$

where $x(e^{j\hat{\omega}_k})$ is a continuous function of frequency, $\hat{\omega}_k$ is a certain frequency sample equaling $\frac{2\pi k}{N}$, and L denotes the length of $x[n]$. This formula analyzes the components of the signal (i.e., the total proportions of various frequencies), as shown in Figure 5. Using this technique, the model can subsequently learn the features of the tool-wear data in the frequency domain.

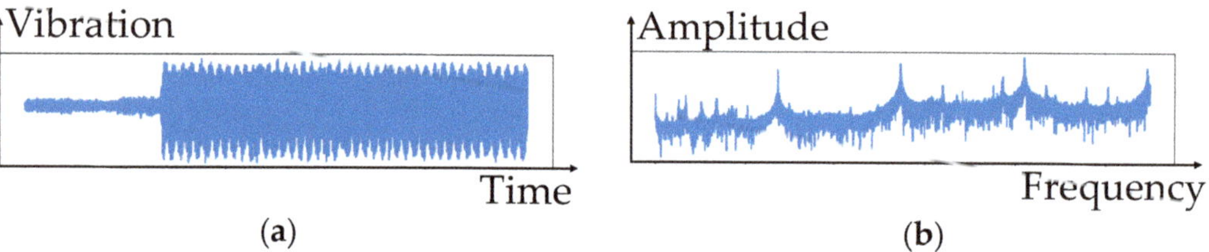

Figure 5. Example of tool-wear vibration data converted into frequency domain using FFT: (**a**) original tool vibration signal; and (**b**) data in frequency domain.

3.3.3. Continuous Wavelet Transform

The continuous wavelet transform technique was first proposed by Grossman et al. [54]. It uses a continuous function to process continuous time data, thereby obtaining a wavelet coefficient to analyze changes in frequency at different times and extensions. Ultimately, the goal of converting the time series data into time–frequency domain data is achieved. Due to space restrictions and the maturity of this technique, we will not go into the details here. Figure 6 displays an example of transformed tool-wear vibration data. This figure clearly shows that the two sections with completely different vibration signals remain different following conversion into a time–frequency graph.

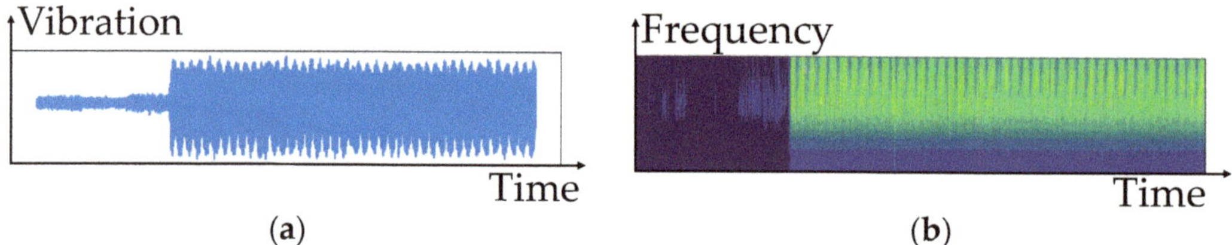

Figure 6. Example of tool-wear vibration data converted into frequency domain using continuous wavelet transform: (**a**) original tool vibration signal; and (**b**) time–frequency graph.

3.4. CNN

We employed a CNN to model the different features. The input of the model can be features extracted using any of the methods discussed in the previous section. The output of the model is the classification of tool-wear status, which can be the rapid initial wear, uniform wear, or failure wear. The framework has one input layer, n convolutional layers, n max pooling layers, and m fully connected layers. The convolutional and max pooling layers are alternated. As the CNN is widely applied, we present only a brief introduction to its framework.

Regarding the input data, suppose that the original data contain n items of data and m features. Thus, the original data will form an $n \times m$ matrix. Applying the feature selection methods to the data then produces h important feature values, from which an $n \times m \times h$ input data matrix can be obtained.

The input layer places the input data in the model. The input data of this layer are the output results, as follows:

$$O_i^{input} = x_i^{input},\tag{13}$$

where O denotes the neuron output, x represents the input data, and i is the ith input of the model.

The purpose of the convolutional layers is to extract each local feature by sliding a filter along the series data. Each node of the neurons in the designed convolutional layers has k kernels of size m. Ultimately, we can write the formula of the neurons as follows:

$$O_i = act\left(\sum_{k=0}^{k=m-1} x_{i+m-k} w_k + b_i\right),\tag{14}$$

where O_i represents the output of the ith neuron, x_i denotes the ith original input, w is the filter, $act(\bullet)$ is an activation function (for which we used Relu), and b_i denotes the bias value of the ith neuron.

Then there are the max pooling layers. Suppose that the input dimensions are (L_i, D_i) and the pooling size and stride of the layers are p and s, respectively. Then, the formula of this layer can be written as follows:

$$L_y = \frac{L_i - p_i}{s} + 1,\tag{15}$$

$$D_y = D_i,\tag{16}$$

and the final output is (L_y, D_y).

The purpose of the fully connected layers is to integrate the outputs of the neurons of the previous layer and then output the classification results. The formula of each neuron is as follows:

$$O_j = act\left(x_j \times w_{ij}\right) + b_j,\tag{17}$$

where O_j represents the output of the jth neuron, x_j is the input of the jth neuron, w_{ij} denotes the weight of the connection with the previous layer, b_j is the bias value of the jth neuron, and $act(\bullet)$ is an activation function. If the fully connected layer is used to output the classification results, then we use the Softmax activation function; if not, then the Relu activation function is used. Finally, the entire model is trained using backpropagation.

Table 4 exhibits the structure of the CNN as applied to the 2010 PHM dataset. In this table, h denotes the number of important features extracted and m is the original number of features. The target CNN contained four sets of convolutional layers and max pooling layers, which would gradually flatten out the important features and thereby enable the extraction of the key factors for classification. We then set the kernel size and stride of the convolutional layers as 3 and 1, respectively, to ensure a full examination of the data. As with most CNN frameworks, we used the Relu activation function in the convolutional layers. Finally, we used two fully connected layers to estimate the output results. As the last layer outputs the classification results, we adopted the Softmax activation function.

Table 4. The structure of the convolutional neural network.

Layer	Type	Output Size	Kernel Size	Stride	Activation Function
Input	Input	$m \times h$	-	-	-
C1	Convolutional	$h \times 2^{\log_2 [m*2]+1}$	3	1	Relu
P2	Max pooling	$(h/2) \times 2^{\log_2 [m*2]+1}$	-	-	-
C3	Convolutional	$(h/2) \times 2^{\log_2 [m*2]+2}$	3	1	Relu
P4	Max pooling	$(h/4) \times 2^{\log_2 [m*2]+2}$	-	-	-
C5	Convolutional	$(h/4) \times 2^{\log_2 [m*2]+3}$	3	1	Relu
P6	Max pooling	$(h/8) \times 2^{\log_2 [m*2]+3}$	-	-	-
C7	Convolutional	$(h/8) \times 2^{\log_2 [m*2]+3}$	3	1	Relu
P8	Max pooling	$(h/16) \times 2^{\log_2 [m*2]+3}$	-	-	-
F9	Fully connected	512	-	-	Relu
Output	Fully connected	3	-	-	softmax

3.5. Use of SNN to Achieve Ensemble Learning

This section introduces the use of an SNN to achieve ensemble learning. First, regarding the input data, suppose that we use a total of h methods to extract the features and that each model has n classification results for tool-wear status. Thus, the input data can be expressed using an $h \times n$ matrix. Then, the input layer places the input data in the model. The input data of this layer are the output results, as follows:

$$O_i^{input} = x_i^{input}, \tag{18}$$

where O denotes the neuron output, x represents the input data, and i is the ith input of the model.

The hidden layer of the SNN comprises multiple fully connected layers, the number of which is determined by the number of previous neurons. Suppose that the output of the previous neuron is α. The fully connected layer has a total of $\log_2 \alpha$ layers. The neurons on each layer integrate the outputs of the neurons in the previous layer, and they are all fully connected. Thus, the formula is as follows:

$$O_j = \left(x_j \times w_{ij}\right) + b_j, \tag{19}$$

where O_j represents the output of the jth neuron, x_j is the input of the jth neuron, w_{ij} denotes the weight of the connection with the previous layer, and b_j is the bias value of the jth neuron.

The objective of the final output layer is to combine the outputs of all the neurons of the previous layer, input them to the activation function, and then output the results. The formula is as follows:

$$O_j = act\left(x_j \times w_{ij}\right) + b_j, \tag{20}$$

where O_j represents the output of the jth neuron, x_j is the input of the jth neuron, w_{ij} denotes the weight of the connection with the previous layer, and b_j is the bias value of the jth neuron. The target problem is a classification problem, so we use the Softmax activation function for $act(\bullet)$.

After modeling, the weights are updated using backpropagation, and the model is trained repeatedly until maximum accuracy is obtained.

4. Experiments

Experiments were conducted to demonstrate the efficiency of the proposed methods. All models and experiments were completed using Python on an Intel Core i7-9700KF CPU at 3.6 GHz with 16 GB member, Nvidia RTX 2080 ti 8 GB GPU, and the Windows 10 operating system.

4.1. Results of Using GAN to Generate Data

This section introduces the parameter settings of the target GAN and the results of using realistic data to complete the original data. First, all training data were normalized using the tool 'minmaxscaler'. We then used the popular tool "Adam" as the optimizer. Through trial and error, we determined that the optimal learning rate was 0.0002. We set the upper limit for epochs at 4000, which takes one day to complete in the selected environment and fits neatly within factory work schedules. For performance evaluation, we referred to Heusel et al. [55] in our use of the Frechet inception distance (FID) score to assess model similarity and the number of iterations to ensure that the distribution of the generated data resembled that of the original data. The FID score calculates the Gaussian distribution distance between the feature vectors of real images and generated images. A smaller value indicates that the Gaussian distribution of the generated image is closer to that of the real image. Figure 7 displays the FID scores during the training process. Observation shows that the FID score was smallest at the 800th epoch, meaning that this distribution was closest to real statuses. Figure 8 compares the generated data under different numbers of epochs. As can be seen, after 800 epochs, the data generated by the GAN model deviated from the original data (Figure 8a), whereas the data generated by the GAN model trained for 800 epochs were similar to the original data. This result corresponds with the FID score results. The results of Figures 7 and 8 indicate that the GAN is subject to overfitting after 800 epochs. Hence, in the following experiments, we trained the model for 800 epochs.

After identifying the optimal parameter settings, we used the GAN to capture rapid initial wear and failure wear data, and the amounts of these two classifications of data were identical to those of uniform wear data. Ultimately, the proportions of original data and GAN-generated data were as shown in Figure 9.

4.2. Validity of Using GAN-Generated Data to Overcome Imbalance in Tool-Wear Data

To verify the validity of using GAN-generated data to overcome data imbalance, we compared the proposed approach with four other methods: (1) directly using the original data without balancing; (2) using augmentation methods to balance the data [56,57]; (3) using SMOTE to balance the data [58–60]; and (4) using downsampling to balance the data [57,61,62]. We compared these five methods by using them to generate a new training set. For the first method, we copied all the original data into the training set. For the second and third methods, we employed the same approach as the proposed GAN method and generated large quantities of rapid initial wear and failure wear data so that the amounts of these two classifications of data equaled that of the uniform wear data. For the final fourth method, with the amount of rapid initial wear data as the benchmark, we randomly extracted the same amount of uniform wear and failure wear data to form the training set. Once the training sets were generated, we used the three feature selection methods (time series feature extraction, FFT, and continuous wavelet transform) to extract important features. The results of each method were used to train a CNN. We therefore trained a total of 15 models (5 methods of handling data imbalance × 3 feature selection methods). Finally, we input the 13,586 items of the original data into these 15 models and observed their prediction results. We used two indices to examine the quality of the prediction results: accuracy and recall. We specifically used recall rather than precision because manufacturers are generally more concerned with the identification of real tool-wear statuses rather than whether each of the predictions is correct.

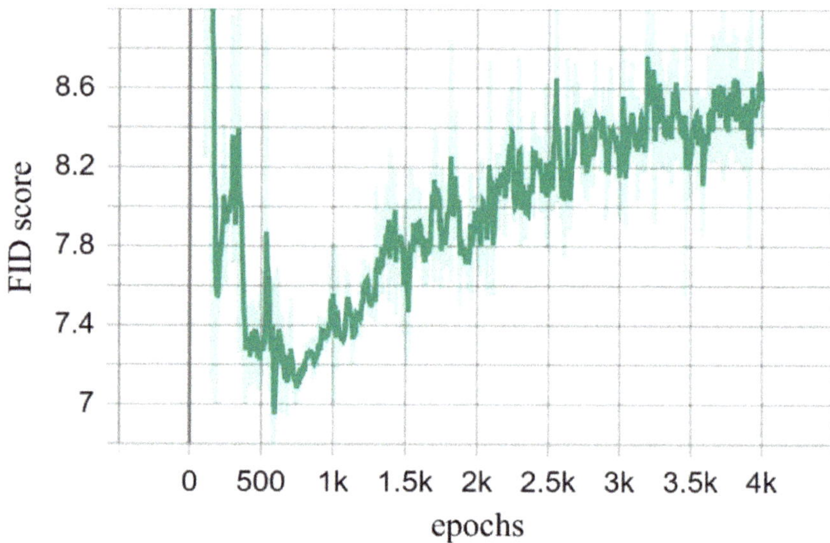

Figure 7. FID scores during training process.

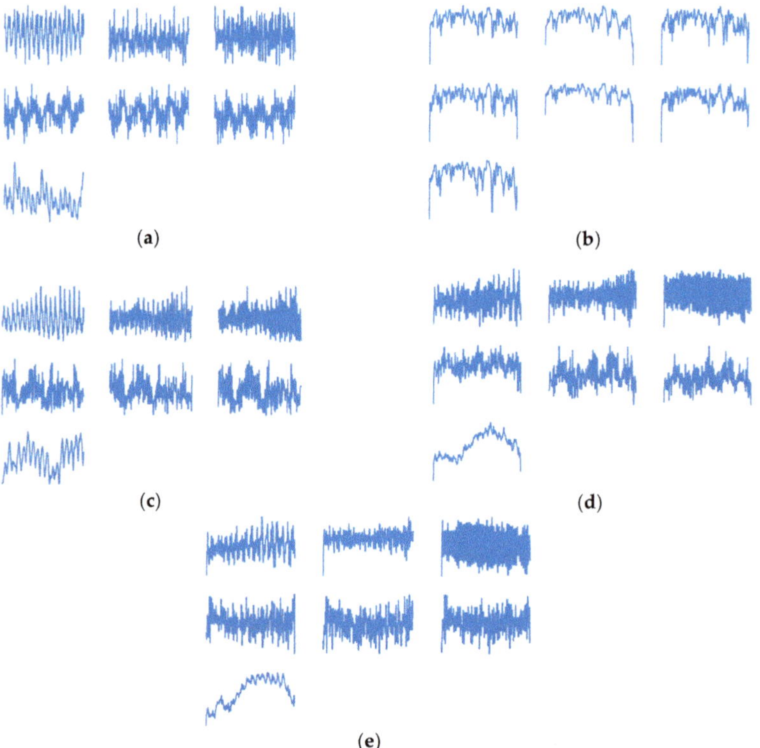

Figure 8. Generated data comparison between different numbers of epochs: (**a**) original data; (**b**) 100 epochs; (**c**) 800 epochs; (**d**) 2000 epochs; and (**e**) 4000 epochs.

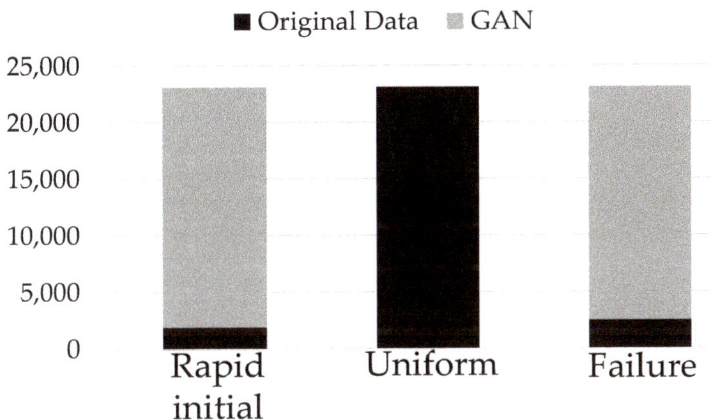

Figure 9. Comparison of data before and after data generation.

Table 5 compares the accuracy values of the 15 models. In this table, we first compared the prediction results of the method using the original data and those of the other methods. Surprisingly, among the methods using time series and FFT, those modeled using the original data had the highest accuracy. Furthermore, among the methods using the continuous wavelet transform, the one modeled using the original data had the third highest accuracy. We speculated that this was because the training data were identical to the test data, and naturally, this led to the highest accuracy. However, in practice, training data would never be identical to test data. Thus, in subsequent analyses, we merely used the prediction data of the methods modeled using the original data as high standards for the other methods (i.e., the prediction results that the models could achieve under the best circumstances). If we apply this standard to assess the prediction performance of the proposed GAN approach, its results come close. This serves as preliminary confirmation of the reasonableness of the proposed approach.

Table 5. Comparison of the accuracy of all combinations of data balancing and feature extraction methods.

	Time Series Feature Extraction	FFT	Continuous Wavelet Transform
Original data	90.08%	88.13%	94.41%
Augmentation	83.40%	84.01%	93.51%
SMOTE	83.45%	82.88%	92.67%
Downsampling	86.09%	86.22%	95.05%
GAN	88.17%	85.85%	96.50%

Then, we compared the prediction performance of the GAN approach with that of the other three data-balancing methods. We found that, regardless of the feature extraction method, the GAN approach provided the most accurate prediction results. This demonstrates that the GAN approach is indeed superior to existing methods in overcoming data imbalance. However, we must mention that the prediction accuracy of the GAN approach paired with FFT was slightly lower than that of the downsampling method. We speculate that this is due to sampling errors in downsampling and the fact that the sample data coincidentally had a distribution similar to that of the test data.

Table 6 compares the recall results of the 15 models with different tool-wear classifications. For the sake of convenience, we presented the amounts of test data for each type of classification. As can be seen, for uniform wear, the recall of the GAN approach was close to 99% and far higher than that of any other method regardless of the feature extraction method. However, for rapid initial wear and failure wear, the recall values were slightly or far lower than those of the other methods. At first glance, these results show that the GAN approach offers no advantages; however, in practice, manufacturers are most interested in the identification of uniform wear status (as stated in the example in Section 1). Rapid initial wear is usually not of concern because it is rare. With regard to failure wear, we can see from Table 6 that the amount of failure wear data is about one-third that of uniform wear, but in practice, tools are generally replaced as soon as failure wear occurs. Thus, there is rarely such a large quantity of failure wear data, which means that manufacturers have little interest in accuracy rates for this classification. Based on the above arguments, we can see that the efficacy of the proposed GAN approach remains valid.

Table 6. Comparison of recall of all combinations of methods with different tool-wear classifications.

	Time Series Feature Extraction			FFT			Continuous Wavelet Transform		
	Rapid Initial	Uniform	Failure	Rapid Initial	Uniform	Failure	Rapid Initial	Uniform	Failure
Number of data	124	9919	3543	124	9919	3543	124	9919	3543
Original data	98.39%	96.22%	72.88%	100.00%	84.50%	98.11%	0.00%	97.43%	89.53%
Augmentation	100.00%	86.95%	73.13%	100.00%	93.30%	57.69%	84.68%	93.93%	92.89%
SMOTE	100.00%	88.03%	70.31%	100.00%	85.35%	75.61%	83.87%	92.48%	93.76%
Downsampling	100.00%	87.68%	81.40%	100.00%	88.05%	80.86%	45.16%	96.85%	92.01%
GAN	97.58%	98.75%	58.51%	100.00%	98.40%	50.52%	64.52%	98.93%	91.08%

4.3. Verification of Necessity of Multiple Feature Extraction Methods for Tool Wear

Most existing studies used a single feature extraction method to predict tool wear. However, we believe that this approach is flawed and therefore used three feature extraction methods for tool-wear prediction. We then combined the prediction results of these methods to derive our final prediction results. We use Table 7 to verify this approach. The table compares the recall results of the three feature extraction methods with the ensemble model.

Table 7. Comparison of recall of different feature extraction methods for different tool-wear statuses.

	Rapid Initial Wear (Recall)	Uniform Wear (Recall)	Failure Wear (Recall)
Time series feature extraction	97.58%	98.75%	58.51%
FFT	100.00%	98.40%	50.52%
Continuous wavelet transform	64.52%	98.93%	91.08%
Ensemble	98%	99%	88%

We first observed the results of the three individual feature extraction methods, which clearly show that the prediction results of time series and FFT were better with regard to rapid initial wear and uniform wear but were poorer with regard to failure wear. We speculate that because these two methods only consider time or frequency information, the features of which do not differ significantly at the junction in uniform wear and failure wear, they could not differentiate between failure wear and uniform wear data. Then, we found that the continuous wavelet transform approach produced better prediction results for uniform wear and failure wear but poor prediction results for rapid initial wear. We believe that this is because the wavelet transform approach extracts the time and frequency features from the data at the same time, which benefits the identification of uniform wear and failure wear data at the junction. However, this also provided too much information and made it difficult to differentiate between the relatively simple rapid initial wear and uniform wear.

Then, comparing the results of the ensemble model with those of the three feature extraction methods, we found that the ensemble model could achieve relatively good accuracy for all three tool-wear statuses. Compared to the three feature extraction methods, which could only obtain superior results for two tool-wear statuses, the results of the ensemble model were significantly better. This demonstrates the validity of using the ensemble approach to integrate different feature extraction methods.

5. Conclusions and Directions for Future Research

Traditionally, manufacturers have relied on experience to determine when a tool should be replaced. While many researchers have developed algorithms to automate this process using CNC machine operating data, most existing studies focus on model development and neglect two fundamental issues in machine learning: data imbalance and feature extraction. In view of this, we applied two approaches for improvement: (1) using a GAN to generate realistic CNC machine vibration data to overcome data imbalance and (2) extracting features in the time, frequency, and time–frequency domains simultaneously and integrating these in an ensemble model. The experimental results demonstrate the validity of the proposed approaches.

In future work, we plan to modify the proposed GAN into a conditional GAN to consider other relevant factors that influence tool wear, such as spindle speed or feed rate, to produce more realistic data.

Author Contributions: Conceptualization, Y.-C.C. (Yi-Chung Chen); Data curation, B.-X.C. and Y.-C.C. (Yi-Chung Chen); Formal analysis, Y.-C.C. (Yi-Chung Chen); Funding acquisition, Y.-C.C. (Yi-Chung Chen) and C.-T.S.; Investigation, Y. C.C. (Yi Chung Chen) and C.-H.L.; Meth-odology, B.-X.C. and Y.-C.C. (Yi-Chung Chen); Project administration, Y.-C.C. (Yi-Chung Chen), C.-H.L., Y.-C.C. (Ying-Chun Chou) and C.-T.S.; Resources, Y.-C.C. (Yi-Chung Chen); Software, B.-X.C.; Supervision, Y.-C.C. (Yi-Chung Chen), C.-H.L., F.-C.W. and C.-T.S.; Validation, B.-X.C., Y.-C.C. (Yi-Chung Chen) and Y.-C.C. (Ying-Chun Chou); Visualization, B.-X.C. and F.-C.W.; Writing—original draft, B.-X.C. and Y.-C.C. (Yi-Chung Chen); Writing—review & editing, Y.-C.C. (Yi-Chung Chen), C.-H.L., Y.-C.C. (Ying-Chun Chou) and F.-C.W. All authors have read and agreed to the published version of the manuscript.

Funding: This study was supported by the Research Assistantships funded by the Ministry of Science and Technology, Taiwan (grant number MOST 108-2622-E-224-014-CC3, MOST 110-2121-M-224-001, MOST 111-2121-M-224-001, to Y.-C.C.).

Data Availability Statement: The research data set of the experiment can be found on 2010 PHM Society Conference Data Challenge. (https://www.phmsociety.org/competition/phm/10, accessed on 1 May 2022).

Conflicts of Interest: The authors declare no conflict of interest.

References

1. Hu, Y.; Miao, X.; Si, Y.; Pan, E.; Zio, E. Prognostics and health management: A review from the perspectives of design, development and decision. *Reliab. Eng. Syst. Saf.* **2022**, *217*, 108063. [CrossRef]
2. Zio, E. Prognostics and Health Management (PHM): Where are we and where do we (need to) go in theory and practice. *Reliab. Eng. Syst. Saf.* **2022**, *218*, 108119. [CrossRef]
3. Vrignat, P.; Kratz, F.; Avila, M. Sustainable manufacturing, maintenance policies, prognostics and health management: A literature review. *Reliab. Eng. Syst. Saf.* **2022**, *218*, 108140. [CrossRef]
4. Sun, C.; Wang, P.; Yan, R.Q.; Gao, R.X.; Chen, X.F. Machine health monitoring based on locally linear embedding with kernel sparse representation for neighborhood optimization. *Mech. Syst. Signal Processing* **2019**, *114*, 25–34. [CrossRef]
5. Sun, C.; Ma, M.; Zhao, Z.; Tian, S.; Yan, R.; Chen, X. Deep transfer learning based on sparse auto-encoder for remaining useful life prediction on tool in manufacturing. *IEEE Trans. Ind. Inform.* **2018**, *15*, 2416–2425. [CrossRef]
6. Chen, B.; Chen, X.; Li, B.; He, Z.; Cao, H.; Cai, G. Reliability estimation for cutting tools based on logistic regression model using vibration signals. *Mech. Syst. Signal Processing* **2011**, *25*, 2526–2537. [CrossRef]
7. Kong, D.; Chen, Y.; Li, N. Gaussian process regression for tool wear prediction. *Mech. Syst. Signal Processing* **2018**, *104*, 556–574. [CrossRef]
8. Benkedjouh, T.; Medjaher, K.; Zerhouni, N.; Rechak, S. Health assessment and life prediction of cutting tools based on support vector regression. *J. Intell. Manuf.* **2015**, *26*, 213–223. [CrossRef]
9. Cai, G.; Chen, X.; Chen, B.L.B.; He, Z. Operation reliability assessment for cutting tools by applying a proportional covariate model to condition monitoring information. *Sensors* **2012**, *12*, 12964–12987. [CrossRef] [PubMed]
10. Zhu, K.; Liu, T. Online Tool Wear Monitoring via Hidden Semi-Markov Model with Dependent Durations. *IEEE Trans. Ind. Inform.* **2018**, *14*, 69–78. [CrossRef]
11. Yu, J.; Liang, S.; Tang, D.; Liu, H. A weighted hidden Markov model approach for continuous-state tool wear monitoring and tool life prediction. *Int. J. Adv. Manuf. Technol.* **2017**, *91*, 201–211. [CrossRef]

12. Kurek, J.; Wieczorek, G.; Kruk, B.S.M.; Jegorowa, A.; Osowski, S. Transfer learning in recognition of drill wear using convolutional neural network. In Proceedings of the 2017 18th International Conference on Computational Problems of Electrical Engineering (CPEE), Kutna Hora, Czech Republic, 11–13 September 2017.
13. Rohan, A. A Holistic Fault Detection and Diagnosis System in Imbalanced, Scarce, Multi-Domain (ISMD) Data Setting for Component-Level Prognostics and Health Management (PHM). *arXiv* **2022**, arXiv:2204.02969. [CrossRef]
14. Zhang, J.; Wang, P.; Yan, R.; Gao, R. Deep Learning for Improved System Remaining Life Prediction. *Procedia CIRP* **2018**, *72*, 1033–1038. [CrossRef]
15. Cao, P.; Zhang, S.; Tan, J. Preprocessing-Free Gear Fault Diagnosis Using Small Datasets with Deep Convolutional Neural Network-Based Transfer Learning. *IEEE Access* **2018**, *6*, 26241–26253. [CrossRef]
16. Chiu, S.M.; Chen, Y.C.; Kuo, C.J.; Hung, L.C.; Hung, M.H.; Chen, C.C.; Lee, C. Development of Lightweight RBF-DRNN and Automated Framework for CNC Tool-Wear Prediction. *IEEE Trans. Instrum. Meas.* **2022**, *71*, 2506711. [CrossRef]
17. Carino, J.A.; Delgado-Prieto, M.; Iglesias, J.A.; Sanchis, A.; Zurita, D.; Millan, M.; Redondo, J.A.O.; Romero-Troncoso, R. Fault Detection and Identification Methodology under an Incremental Learning Framework Applied to Industrial Machinery. *IEEE Access* **2018**, *6*, 49755–49766. [CrossRef]
18. Brito, L.C.; da Silva, M.B.; Duarte, M.A.V. Identification of cutting tool wear condition in turning using self-organizing map trained with imbalanced data. *J. Intell. Manuf.* **2021**, *32*, 127–140. [CrossRef]
19. Miao, H.; Zhao, Z.; Sun, C.; Li, B.; Yan, R. A U-Net-Based Approach for Tool Wear Area Detection and Identification. *IEEE Trans. Instrum. Meas.* **2021**, *70*, 5004110. [CrossRef]
20. Rohan, A.; Raouf, I.; Kim, H.S. Rotate Vector (RV) Reducer Fault Detection and Diagnosis System: Towards Component Level Prognostics and Health Management (PHM). *Sensors* **2020**, *20*, 6845. [CrossRef]
21. Goodfellow, I.; Pouget-Abadie, J.; Mirza, M.; Xu, B.; Warde-Farley, D.; Ozair, S.; Aaron, C.; Benglo, Y. Generative adversarial nets. *Adv. Neural Inf. Processing Syst.* **2014**, *27*, 2672–2680.
22. Karras, T.; Aila, T.; Laine, S.; Lehtinen, J. Progressive Growing of GANs for Improved Quality, Stability, and Variation. *arXiv* **2017**, arXiv:1710.10196.
23. Yadav, N.K.; Singh, S.K.; Dubey, S.R. CSA-GAN: Cyclic synthesized attention guided generative adversarial network for face synthesis. *Appl. Intell.* **2022**. [CrossRef]
24. Shi, Y.; Aggarwal, D.; Jain, A.K. Lifting 2D StyleGAN for 3D-Aware Face Generation. In Proceedings of the IEEE/CVF Conference on Computer Vision and Pattern Recognition (CVPR), Nashville, TN, USA, 20–25 June 2021; pp. 6258–6266.
25. Fang, Z.; Liu, Z.; Liu, T.; Hung, C.C.; Xiao, J.; Feng, G. Facial expression GAN for voice-driven face generation. *Vis. Comput.* **2022**, *38*, 1151–1164. [CrossRef]
26. Chen, Y.; Zhang, H.; Liu, L.; Chen, X.; Zhang, Q.; Yang, K.; Xia, R.; Xie, J. Research on image Inpainting algorithm of improved GAN based on two-discriminations networks. *Appl. Intell.* **2021**, *51*, 3460–3474. [CrossRef]
27. Wei, D.; Huang, K.; Ma, L.; Hua, J.; Lai, B.; Shen, H. OAW-GAN: Occlusion-aware warping GAN for unified human video synthesis. *Appl. Intell.* **2022**, 1–18. [CrossRef]
28. Tagawa, Y.; Maskeliūnas, R.; Damaševičius, R. Acoustic Anomaly Detection of Mechanical Failures in Noisy Real-Life Factory Environments. *Electronics* **2021**, *10*, 2329. [CrossRef]
29. Zhang, F.; Ma, Y.; Yuan, G.; Zhang, H.; Ren, J. Multiview image generation for vehicle reidentification. *Appl. Intell.* **2021**, *51*, 5665–5682. [CrossRef]
30. Gan, Y.S.; Liong, S.-T.; Wang, S.-Y.; Cheng, C.T. An improved automatic defect identification system on natural leather via generative adversarial network. *Int. J. Comput. Integr. Manuf.* **2022**, 1–17. [CrossRef]
31. Gu, J.; Qi, Y.; Zhao, Z.; Su, W.; Su, L.; Li, K.; Pecht, M. Fault diagnosis of rolling bearings based on generative adversarial network and convolutional denoising auto-encoder. *J. Adv. Manuf. Sci. Technol.* **2022**, *2*, 2022009. [CrossRef]
32. Wang, M.; Zhou, J.; Gao, J.; Li, Z.; Li, E. Milling Tool Wear Prediction Method Based on Deep Learning under Variable Working Conditions. *IEEE Access* **2020**, *8*, 140726–140735. [CrossRef]
33. *ISO 3685:1993*; Tool-Life Testing with Single-point Turning Tools. International Organization for Standardization: Geneva, Switzerland, 1993.
34. Ertürk, Ş.; Kayabaşi, O. Investigation of the Cutting Performance of Cutting Tools Coated with the Thermo-Reactive Diffusion (TRD) Technique. *IEEE Access* **2019**, *7*, 106824–106838. [CrossRef]
35. Bhuiyan, M.S.H.; Choudhury, I.A.; Dahari, M.; Nukman, Y. Application of acoustic emission sensor to investigate the frequency of tool wear and plastic deformation in tool condition monitoring. *Measurement* **2016**, *92*, 208–217. [CrossRef]
36. Dolinšek, S.; Kopač, J. Acoustic emission signals for tool wear identification. *Wear* **1999**, *225*, 295–303. [CrossRef]
37. Bhuiyan, M.S.H.; Choudhury, I.A.; Nukma, Y.n. An innovative approach to monitor the chip formation effect on tool state using acoustic emission in turning. *Int. J. Mach. Tools Manuf.* **2012**, *58*, 19–28.
38. Li, Z.; Liu, R.; Wu, D. Data-driven smart manufacturing: Tool wear monitoring with audio signals and machine learning. *J. Manuf. Processes* **2019**, *48*, 66–76. [CrossRef]
39. Gomes, M.C.; Brito, L.C.; da Silva, M.B.; Duarte, M.A.V. Tool wear monitoring in micromilling using Support Vector Machine with vibration and sound sensors. *Precis. Eng.* **2021**, *67*, 137–151. [CrossRef]

40. Mohanraj, T.; Yerchuru, J.; Krishnan, H.; Aravind, R.S.N.; Yameni, R. Development of tool condition monitoring system in end milling process using wavelet features and Hoelder's exponent with machine learning algorithms. *Measurement* **2021**, *173*, 108671. [CrossRef]
41. Jalali, S.K.; Ghandi, H.; Motamedi, M. Intelligent condition monitoring of ball bearings faults by combination of genetic algorithm and support vector machines. *J. Nondestruct. Eval.* **2020**, *39*, 25. [CrossRef]
42. Corne, R.; Nath, C.; el Mansori, M.; Kurfess, T. Study of spindle power data with neural network for predicting real-time tool wear/breakage during inconel drilling. *J. Manuf. Syst.* **2017**, *43*, 287–295. [CrossRef]
43. Hesser, D.F.; Markert, B. Tool wear monitoring of a retrofitted CNC milling machine using artificial neural networks. *Manuf. Lett.* **2019**, *19*, 1–4. [CrossRef]
44. Zhao, R.; Yan, R.; Wang, J.; Mao, K. Learning to monitor machine health with convolutional bi-directional LSTM networks. *Sensors* **2017**, *17*, 273. [CrossRef] [PubMed]
45. Cheng, C.; Li, J.; Liu, Y.; Nie, M.; Wang, W. Deep convolutional neural network-based in-process tool condition monitoring in abrasive belt grinding. *Comput. Ind.* **2019**, *106*, 1–13. [CrossRef]
46. Zhao, R.; Wang, D.; Yan, R.; Mao, K.; Shen, F.; Wang, J. Machine health monitoring using local feature-based gated recurrent unit networks. *IEEE Trans. Ind. Electron.* **2017**, *65*, 1539–1548. [CrossRef]
47. 2010 PHM Society Conference Data Challenge. Available online: https://www.phmsociety.org/competition/phm/10 (accessed on 1 June 2022).
48. Chen, Y.C.; Li, D.C. Selection of key features for PM2.5 prediction using a wavelet model and RBF-LSTM. *Appl. Intell.* **2021**, *51*, 2534–2555. [CrossRef]
49. Kuo, C.J.; Ting, K.C.; Chen, Y.C.; Yang, D.L.; Chen, H.M. Automatic machine status prediction in the era of industry 4.0: Case study of machines in a spring factory. *J. Syst. Archit.* **2017**, *81*, 44–53. [CrossRef]
50. Duerden, A.; Marshall, F.E.; Moon, N.; Swanson, C.; Donnell, K.; Grubbs, G.S., II. A chirped pulse Fourier transform microwave spectrometer with multi-antenna detection. *J. Mol. Spectrosc.* **2021**, *376*, 111396. [CrossRef]
51. Xu, W.; Xu, K.J.; Yu, X.L.; Huang, Y.; Wu, W.K. Signal processing method of bubble detection in sodium flow based on inverse Fourier transform to calculate energy ratio. *Nucl. Eng. Technol.* **2021**, *53*, 3122–3125. [CrossRef]
52. Jalayer, M.; Orsenigo, C.; Vercellis, C. Fault detection and diagnosis for rotating machinery: A model based on convolutional LSTM, Fast Fourier and continuous wavelet transforms. *Comput. Ind.* **2021**, *125*, 103378. [CrossRef]
53. Koga, K. Signal processing approach to mesh refinement in simulations of axisymmetric droplet dynamics. *J. Comput. Appl. Math.* **2021**, *383*, 113131. [CrossRef]
54. Grossmann, A.; Kronland-Martinet, R.; Morlet, J. Reading and Understanding Continuous Wavelet Transforms. In *Wavelets. Inverse Problems and Theoretical Imaging*; Combes, J.M., Grossmann, A., Tchamitchian, P., Eds.; Springer: Berlin/Heidelberg, Germany, 1989.
55. Heusel, M.; Ramsauer, H.; Unterthiner, T.; Nessler, B.; Hochreiter, S. GANs trained by a two time-scale update rule converge to a local nash equilibrium. In Proceedings of the International Conference on Neural Information Processing Systems (NIPS'17), Long Beach, CA, USA, 4–9 December 2017; pp. 6629–6640.
56. Shorten, C.; Khoshgoftaar, T.M. A survey on image data augmentation for deep learning. *J. Big Data* **2019**, *6*, 1–48. [CrossRef]
57. Koziarski, M.; Kwolek, B.; Cyganek, B. Convolutional neural network-based classification of histopathological images affected by data imbalance. In Proceedings of the Video Analytics. Face and Facial Expression Recognition, 3rd International Workshop, FFER 2018, and 2nd International Workshop, DLPR 2018, Beijing, China, 20 August 2018; pp. 1–11.
58. Bustillo, A.; Rodríguez, J.J. Online breakage detection of multitooth tools using classifier ensembles for imbalanced data. *Int. J. Syst. Sci.* **2014**, *45*, 2590–2602. [CrossRef]
59. Mathew, J.; Pang, C.K.; Luo, M.; Leong, W.H. Classification of imbalanced data by oversampling in kernel space of support vector machines. *IEEE Trans. Neural Netw. Learn. Syst.* **2017**, *29*, 4065–4076. [CrossRef] [PubMed]
60. Chawla, N.V.; Bowyer, K.W.; Hall, L.O.; Kegelmeyer, W.P. SMOTE: Synthetic minority over-sampling technique. *J. Artif. Intell. Res.* **2002**, *16*, 321–357. [CrossRef]
61. Hassan, A.R.; Bhuiyan, M.I.H. Automated identification of sleep states from EEG signals by means of ensemble empirical mode decomposition and random under sampling boosting. *Comput. Methods Programs Biomed.* **2017**, *140*, 201–210. [CrossRef] [PubMed]
62. Krawczyk, B.; Galar, M.; Jeleń, Ł.; Herrera, F. Evolutionary undersampling boosting for imbalanced classification of breast cancer malignancy. *Appl. Soft Comput.* **2016**, *38*, 714–726. [CrossRef]

 *electronics*

Article

Classifying Conditions of Speckle and Wrinkle on the Human Face: A Deep Learning Approach

Tsai-Rong Chang * and Ming-Yen Tsai

Department of Computer Science and Information Engineering, Southern Taiwan University of Science and Technology, No. 1, Nantai Street, Yongkang District, Tainan 71005, Taiwan
* Correspondence: trchang@stust.edu.tw

Abstract: Speckles and wrinkles are common skin conditions on the face, with occurrence ranging from mild to severe, affecting an individual in various ways. In this study, we aim to detect these conditions using an intelligent deep learning approach. First, we applied a face detection model and identified the face image using face positioning techniques. We then split the face into three polygonal areas (forehead, eyes, and cheeks) based on 81 position points. Skin conditions in the images were firstly judged by skin experts and subjectively classified into different categories, from good to bad. Wrinkles were classified into five categories, and speckles were classified into four categories. Next, data augmentation was performed using the following manipulations: changing the HSV hue, image rotation, and horizontal flipping of the original image, in order to facilitate deep learning using the Resnet models. We tested the training using these models each with a different number of layers: ResNet-18, ResNet-34, ResNet-50, ResNet-101, and ResNet-152. Finally, the K-fold (K = 10) cross-validation process was applied to obtain more rigorous results. Results of the classification are, in general, satisfactory. When compared across models and across skin features, we found that Resnet performance is generally better in terms of average classification accuracy when its architecture has more layers.

Keywords: image processing; skin condition detection; deep learning; ResNet

Citation: Chang, T.-R.; Tsai, M.-Y. Classifying Conditions of Speckle and Wrinkle on the Human Face: A Deep Learning Approach. *Electronics* **2022**, *11*, 3623. https://doi.org/10.3390/electronics11213623

Academic Editor: Chiman Kwan

Received: 30 August 2022
Accepted: 3 November 2022
Published: 6 November 2022

1. Introduction

Speckles and wrinkles on the face are common skin conditions. They affect an individual's appearance to different degrees. Wrinkles typically increase with age. With skin care [1], the appearance of such skin features may be delayed. Skin conditions, when serious, are typically treated by clinicians after visual assessment, though they are often difficult to diagnose with the naked eye. Here, we aim to implement a diagnostic tool for clinicians to facilitate their professional assessments, focusing on classifying grades of skin conditions.

Skin is the largest organ in the human body [2]. It is divided into three parts: epidermis, dermis, and subcutaneous tissue (Figure 1).

Physicians routinely examine skin abnormality using the non-invasive dermatoscope with optical magnification. Some skin diseases are misdiagnosed, even by professional doctors [3]. To improve subjective judgment, images obtained by the dermatoscope have been subjected to deep learning models [4]. Lesions from other systemic diseases [5–8], and oral diseases have been similarly identified through deep learning [9].

In 2004, Mukaida et al. [10] proposed to extract wrinkles and spots based on local analysis of their shape characteristics. In 2015, Ng et al. [11] proposed the method of Hessian line tracking (HLT), based on the Hessian filter. This method strengthens connectivity of wrinkles, improving the accuracy of locating wrinkles. In 2017, Canak et al. [12] used local binary patterns to extract wrinkle features. In 2017, Zaghbani et al. [13] used the Gabor filter to extract wrinkles for facial emotion studies. These methods are used exclusively to detect wrinkles and spots, but do not distinguish the severity or grading of the skin conditions.

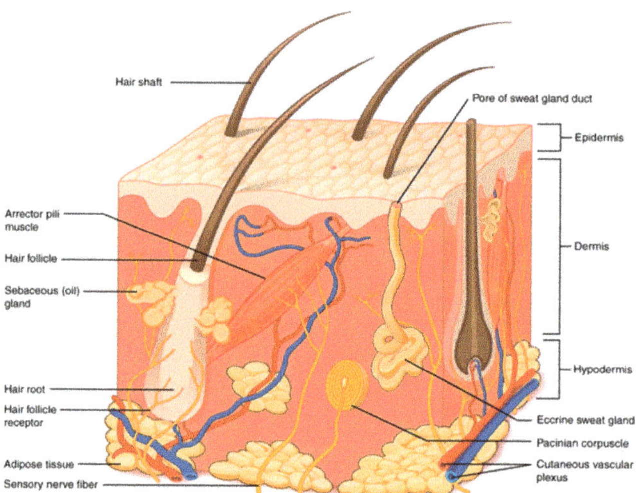

Figure 1. Diagram of skin anatomy [2].

In face detection, several methods have been proposed, including OpenCV's Haar Cascade Classifier, Dlib's Histogram of Gradient (HOG), and Dlib's convolutional neural network (CNN). The object-detection in OpenCV is Cascade Classifier, which is a boosting method serially connecting multiple weak classifiers. The early Cascade Classifier of OpenCV makes use of Haar-like features, later adding Local Binary Pattern (LBP) and Histogram of Gradient (HOG). In 2001, Viola and Jones proposed an algorithm of Viola–Jones' Object detection framework. Face feature-point detection is used to identify common features on the face [14], by comparing dark circles and upper cheeks, brighter area of the nose-bridge between two eyes, and the nose-bridge, and by detecting feature point positions of the eye, nose and mouth.

In recent years, applications of face images have been growing. For example, the COVID-19 epidemic has forced people to wear face masks. Other applications include popular software for special effects such as processing of faces, fast face synthesis sketching [15], and face swapping [16]. In 2017, Rosebrock proposed a 68-point face model for multi-point image capture for face alignment [17], as shown in Figure 2a, based on the Dlib library. Through this algorithm, facial features are first marked with numbers, and then the target block is intercepted according to these points. With this approach, face images can be used for training by deep learning. Subsequently, various applications have been achieved: e.g., face changing [16], head pose estimation [18], expression discrimination [19], and emotion analysis [20]. In 2019, other researchers added 13 to the 68 points to form an 81-point model [21], further improving the performance.

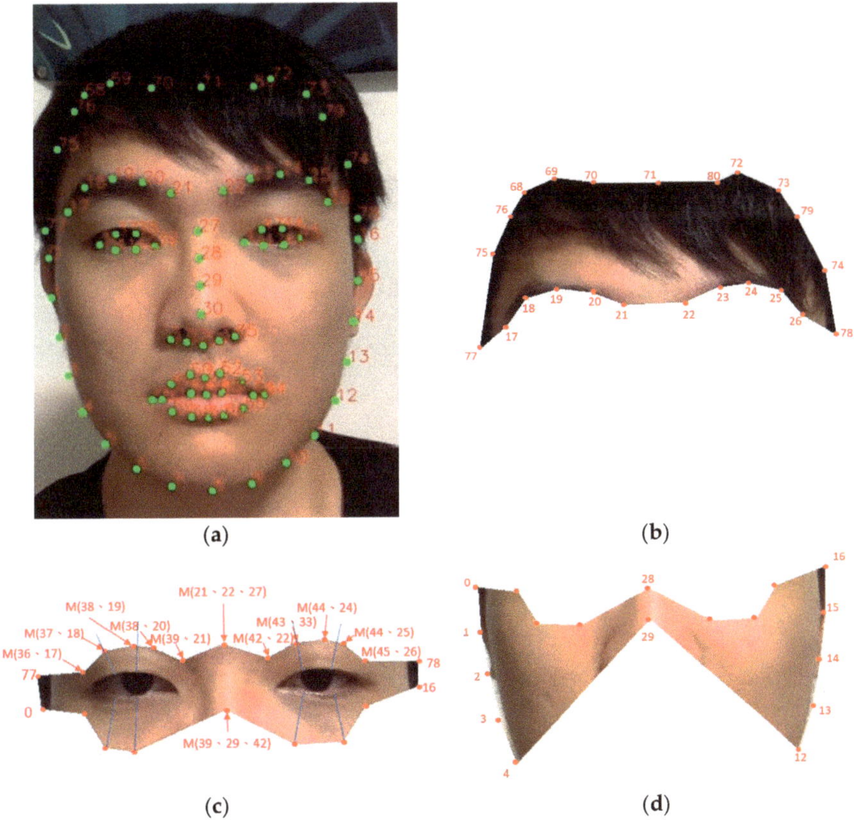

Figure 2. (**a**) A result plot produced by the 81-point model showing the point-locations in green; (**b**) Defined forehead region extracted from (**a**); (**c**) Defined eyes region; and (**d**) Defined cheeks region also extracted from (**a**).

2. Materials and Methods

Deep learning is gradually becoming integrated into everyday life. Through deep learning, computers can learn and may even surpass human performance. In this study, we applied a supervised system for feature detection. For relevant training, the system was given multiple pictures with answer labels. These answer labels were produced by human experts. To generate classified results consistent with the perspectives of both the public and experts, we invited professional beauticians and non-professional people to help label the skin conditions in the images, with slightly more weight given to the professionals.

In order to isolate facial regions, we used Dlib's 81-point model for detecting facial feature points and for their polygonal segmentation. To improve the image processing, some technical issues needed to be resolved regarding image capturing. These issues included sufficient lighting, and camera stability using a tripod.

We first applied a facial positioning technique of 81-points to divide the face into three gross areas (forehead, eyes, and cheeks). We then increased the original number of images through data augmentation. Data augmentation involved the following manipulations: changing the HSV, and changing the rotational angle of the original image to produce new images. These data were then trained via the residual neural network (ResNet). Since there is only one result after training, it is necessary to use K-fold (K = 10) cross validation in order to obtain more rigorous results. In K-fold cross validation, 10% of data were used only for testing, not for training. The original data were hence segregated randomly into ten equal parts, and training was repeated ten times at the ratio of 9:1. For each 1000 epochs of training, we also checked results with K-fold (K = 10) cross validation. Whenever the accuracy rate appeared to grow rapidly, training was terminated. This step was to prevent model overfitting.

2.1. Face Detection

Facial landmarks are useful in predicting shapes. In the input image, we judged and assigned areas of interest, based on the shape predictors and the locations of all points of interest along the shape outline. A two-step procedure was followed:

1. Locating the entire face in the image: this step can be achieved in many ways. Three built-in training detectors are available in OpenCV: Haar, LBP, and HOG. Haar feature detection is most popular. Here, we applied the mainstream deep learning algorithm for face positioning. Once the face in the image was detected and the image positioned, we proceeded to the next step.
2. Detecting areas of interest of the face: during the first step, the (x, y) coordinates of the face were obtained, and various facial marker detectors were then applied. Specifically, the following face areas were marked: right eyebrow, left eyebrow, right eye, left eye, nose, mouth, and jaw. By importing the image file in '.dat' format, and using Dlib as the predictor, we were able to display the desired points on the face and the cropped rectangular or polygonal areas from the image.

2.2. Polygon Images

To avoid irrelevant parts affecting training, the image was cut through those 81 points to form polygons each representing areas of forehead, eyes, or cheeks. Figure 2a shows an example of the 81-point model. We further defined intermediate points. If A and B are two of the 81 feature points, the intermediate point *M(A, B)* is the mid-point between them as represented by:

$$M(A, B) = round\left(\frac{A_x + B_x}{2}, \frac{A_y + B_y}{2}\right) \tag{1}$$

Sometimes it is necessary to involve three instead of two points according to the following:

$$M(A, B, C) = round\left(\frac{A_x + B_x + C_x}{3}, \frac{A_y + B_y + C_y}{3}\right) \tag{2}$$

2.3. Residual Neural Network (ResNet)

Residual Neural Network (ResNet) was proposed by He et al. in 2016 [22]. Figure 3 shows the various architectures of the ResNet network including: ResNet-18, ResNet-34, ResNet-50, and ResNet-101. Their difference is in the number of layers. Taking the example of ResNet-152, there is a layer of input at the front, plus the middle building block, 3 + 8 + 36 + 3 = 50, and then each block is divided into three layers, giving 50 × 3 = 150, plus the last classification layer. It therefore has a total of 1 + 150 + 1 = 152 layers. In addition, the training parameters used in this study are: number of epochs: 1000, batch size: 128, learning rate: 0.001, momentum: 0.9, optimizer: stochastic gradient descent (SGD), loss function: cross entropy.

2.4. Data Augmentation

In the event of insufficient data, deep learning training will likely fail. It is necessary to augment the amount of data without changing the parameters that affect training. For example, in training for skin color, image volume can be increased by rotating the angle of an image (1 to 5° clockwise or anti-clockwise). This approach is not applicable in training for wrinkles, because the training will be affected. For wrinkles or speckles, we increased the data volume by changing the H and S in HSV, as this does not affect training.

Layer name (output size)	ResNet-18	ResNet-34	ResNet-50	ResNet-101	ResNet-152
Conv1 (112 × 112)	7×7, 64, stride 2				
	3×3 max pool, stride 2				
Conv2_x (56 × 56)	$\begin{bmatrix} 3 \times 3, 64 \\ 3 \times 3, 64 \end{bmatrix} \times 2$	$\begin{bmatrix} 3 \times 3, 64 \\ 3 \times 3, 64 \end{bmatrix} \times 3$	$\begin{bmatrix} 1 \times 1, 64 \\ 3 \times 3, 64 \\ 1 \times 1, 256 \end{bmatrix} \times 3$	$\begin{bmatrix} 1 \times 1, 64 \\ 3 \times 3, 64 \\ 1 \times 1, 256 \end{bmatrix} \times 3$	$\begin{bmatrix} 1 \times 1, 64 \\ 3 \times 3, 64 \\ 1 \times 1, 256 \end{bmatrix} \times 3$
Conv3_x (28 × 28)	$\begin{bmatrix} 3 \times 3, 128 \\ 3 \times 3, 128 \end{bmatrix} \times 2$	$\begin{bmatrix} 3 \times 3, 128 \\ 3 \times 3, 128 \end{bmatrix} \times 4$	$\begin{bmatrix} 1 \times 1, 128 \\ 3 \times 3, 128 \\ 1 \times 1, 512 \end{bmatrix} \times 4$	$\begin{bmatrix} 1 \times 1, 128 \\ 3 \times 3, 128 \\ 1 \times 1, 512 \end{bmatrix} \times 4$	$\begin{bmatrix} 1 \times 1, 128 \\ 3 \times 3, 128 \\ 1 \times 1, 512 \end{bmatrix} \times 8$
Conv4_x (14 × 14)	$\begin{bmatrix} 3 \times 3, 256 \\ 3 \times 3, 256 \end{bmatrix} \times 2$	$\begin{bmatrix} 3 \times 3, 256 \\ 3 \times 3, 256 \end{bmatrix} \times 6$	$\begin{bmatrix} 1 \times 1, 256 \\ 3 \times 3, 256 \\ 1 \times 1, 1024 \end{bmatrix} \times 6$	$\begin{bmatrix} 1 \times 1, 256 \\ 3 \times 3, 256 \\ 1 \times 1, 1024 \end{bmatrix} \times 23$	$\begin{bmatrix} 1 \times 1, 256 \\ 3 \times 3, 256 \\ 1 \times 1, 1024 \end{bmatrix} \times 36$
Conv5_x (7 × 7)	$\begin{bmatrix} 3 \times 3, 512 \\ 3 \times 3, 512 \end{bmatrix} \times 2$	$\begin{bmatrix} 3 \times 3, 512 \\ 3 \times 3, 512 \end{bmatrix} \times 3$	$\begin{bmatrix} 1 \times 1, 512 \\ 3 \times 3, 512 \\ 1 \times 1, 2048 \end{bmatrix} \times 3$	$\begin{bmatrix} 1 \times 1, 512 \\ 3 \times 3, 512 \\ 1 \times 1, 2048 \end{bmatrix} \times 3$	$\begin{bmatrix} 1 \times 1, 512 \\ 3 \times 3, 512 \\ 1 \times 1, 2048 \end{bmatrix} \times 3$
FC (1 × 1)	Average pool, 1000-d Full Connected, Softmax				

Figure 3. Details of each layer in the ResNet architecture [22].

Through changing the HSV and rotation angle of the image, and with various combinations, we increased 200 raw images to >10,000 images for training. Specifically, the hue of the image is subjected to H±0 to 5 for a total of 11 amplifications. The altered images did not appear very different to the naked eye. Regarding the change to saturation of the image, we applied five amplifications at S±0 to 2, finally giving $11 \times 5 - 1 = 54$ permutations and combinations. In addition, there are another two types of amplifications with horizontal flips, and six types of amplifications with three rotational angles (left or right by 5, 10, and 15°). Finally, data augmentation increased the original 200 images to $200 \times (11 \times 5 - 1) \times 2 \times 6 = 129{,}600$ images.

3. Results

We analyzed 129,600 facial images obtained after data augmentation. Based on the 81-point model, the face is segmented into three regions: forehead, eyes, and cheeks. Images of the whole face and the segmented areas were first viewed by professional beauticians to score wrinkles and speckles in terms of quality grading. After completing a model's training, we used K-fold (K = 10) cross validation to show the model's performance (namely, accuracy in classifying skin conditions in terms of gradings). Note that in the K-fold cross validation, for each iteration, we used 90% original data plus their augmented data for training, with the remaining 10% (including original and their augmented data) being used for testing and not for training.

3.1. Dataset

The data used in this study is a dataset obtained by collaborating with a commercial company. There is a total of 200 face images. All subjects in the images provided written informed consent (in Chinese) after a clear explanation of the study protocol. As facial features of individual participants were extracted in fragments and randomly coded, facial data were hence not traceable to individuals. There was no privacy issue involved. At the time of taking these images, there was no lighting problems (too bright or too dim). For subsequent training and pre-treatment, we adopted the influence of elimination. Then, we solved the problem of over-reconciliation caused by insufficient data volume, that is, the number of images was increased to 129,600 through data augmentation as described above.

3.2. Skin Grading

To avoid visual fatigue of viewers, images were rated in batches of 20 images each time (from the 200 original images), resulting in ten rating sessions for each viewer. Finally, these ratings were averaged across viewers to become the rating of each image. The images were then augmented as described above. In rating skin conditions, wrinkles have five grades: Grade 1 is the best, and Grade 5 is the worst. For speckles, because they are less common than wrinkles, the sample size is smaller, and there are only four grades. Section 3.2.4 below shows an example of the image database.

3.2.1. Forehead Wrinkles

Figure 4 shows that Grade 1 has almost no wrinkles with obvious skin luster. This grade is found with most young people aged between 20 and 30 years old. With good skin care, some 40 years old also have this skin grade. Grade 2 shows a few fine lines. Grade 4 has obvious wrinkles, but these are not as deep and abundant as in Grade 5. Wrinkles of Grades 2 and 3 are age-related and are found in people aged between 40 to 60 years old. Grades 4 and 5 are common in the elderly.

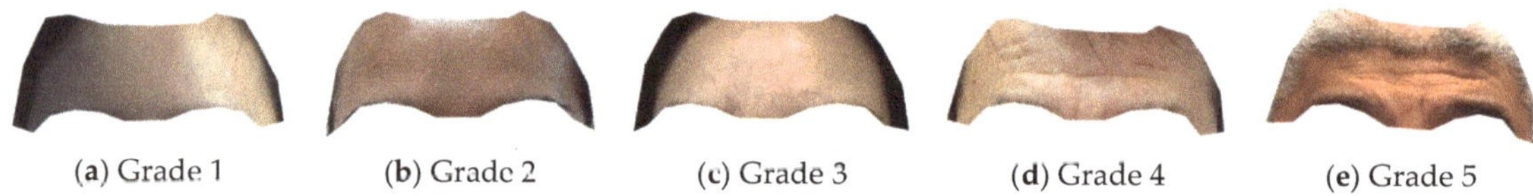

(**a**) Grade 1   (**b**) Grade 2   (**c**) Grade 3   (**d**) Grade 4   (**e**) Grade 5

Figure 4. Examples of the five grades of forehead wrinkles (regions are similar to Figure 2b).

3.2.2. Eye Wrinkles

Figure 5 shows that Grade 1 has no wrinkles around the eyes. Grade 2 has some faint wrinkles at the lower eyelid. Grade 3 has more wrinkles. In Grade 4 the wrinkles appear more obviously, and appear even on the nose. In Grade 5, wrinkles appear deeper. Grades 3 to 5 are age-related, and are found in subjects >40 years of age.

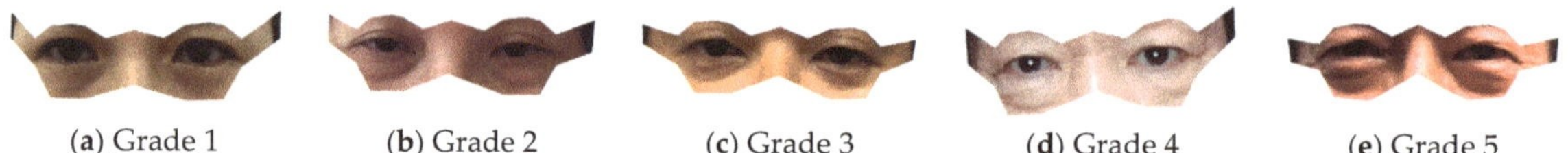

(**a**) Grade 1   (**b**) Grade 2   (**c**) Grade 3   (**d**) Grade 4   (**e**) Grade 5

Figure 5. Examples of the five grades of eye wrinkles (regions are similar to Figure 2c).

3.2.3. Cheek Wrinkles

Although there are some red speckles on cheeks, this is also a classification of wrinkles. Thus, only wrinkles are classified as shown in Figure 6.

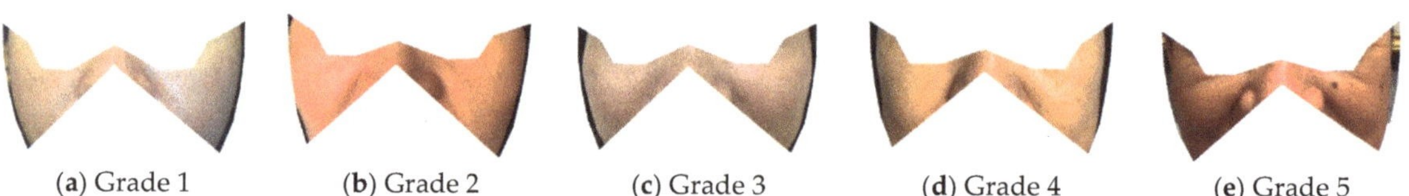

(**a**) Grade 1   (**b**) Grade 2   (**c**) Grade 3   (**d**) Grade 4   (**e**) Grade 5

Figure 6. Examples of the five grades of cheek wrinkles (regions are similar to Figure 2d).

3.2.4. Cheek Speckles

Regarding speckles, only four categories are classified. Grade 1 has very few speckles. Grade 2 has one or two speckles. Grade 3 has a small number of speckles. Grade 4 has even more speckles, commonly in middle-age or older subjects (Figure 7).

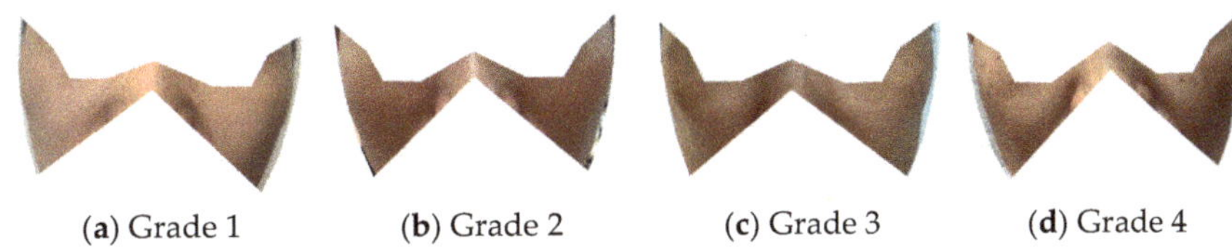

(**a**) Grade 1   (**b**) Grade 2   (**c**) Grade 3   (**d**) Grade 4

Figure 7. Examples of the four grades of cheek speckles (regions are similar to Figure 2d).

3.3. Training Results

3.3.1. Training Results with Polygon Images

Figure 8 shows an example of Grade 1 forehead wrinkles, and the training results using various models. This type of image was originally classified as Grade 1. Through training with ResNet-18, ResNet-34 and ResNet-50, it was classified as Grade 2, while with ResNet-101 and ResNet-152 it is correctly classified as Grade 1. Here, ResNet-152 shows the biggest improvement in accuracy. The poorer performances by models with fewer layers are likely due to presence of hair bangs in the images.

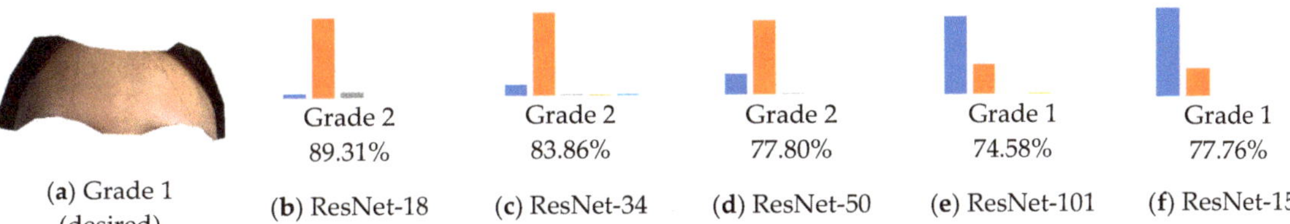

(**a**) Grade 1 (desired)

(**b**) ResNet-18 — Grade 2 89.31%

(**c**) ResNet-34 — Grade 2 83.86%

(**d**) ResNet-50 — Grade 2 77.80%

(**e**) ResNet-101 — Grade 1 74.58%

(**f**) ResNet-151 — Grade 1 77.76%

Figure 8. Example and training results of Grade 1 forehead wrinkles (wrongly classified grades are underlined).

Figure 9 similarly shows Grade 2 eye wrinkles. In ResNet, several architecture models show the correct classifications rather satisfactorily.

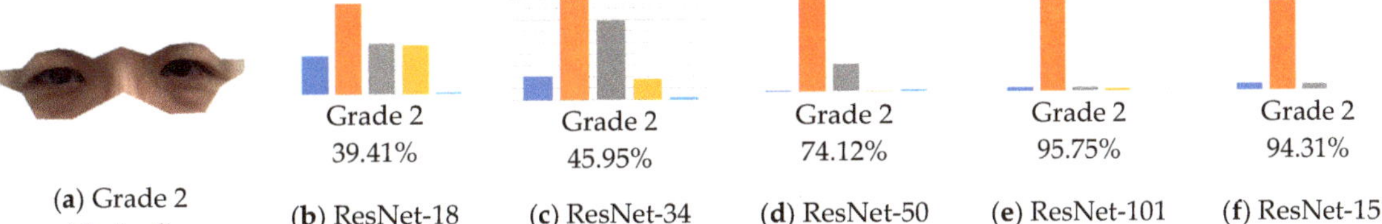

(**a**) Grade 2 (desired)

(**b**) ResNet-18 — Grade 2 39.41%

(**c**) ResNet-34 — Grade 2 45.95%

(**d**) ResNet-50 — Grade 2 74.12%

(**e**) ResNet-101 — Grade 2 95.75%

(**f**) ResNet-151 — Grade 2 94.31%

Figure 9. Example and training results of Grade 2 eye wrinkles.

Figure 10 shows Grade 3 cheek wrinkles. This type of images is not classified correctly even with ResNet-152. Shadows on both sides of the nose likely cause such poor performances.

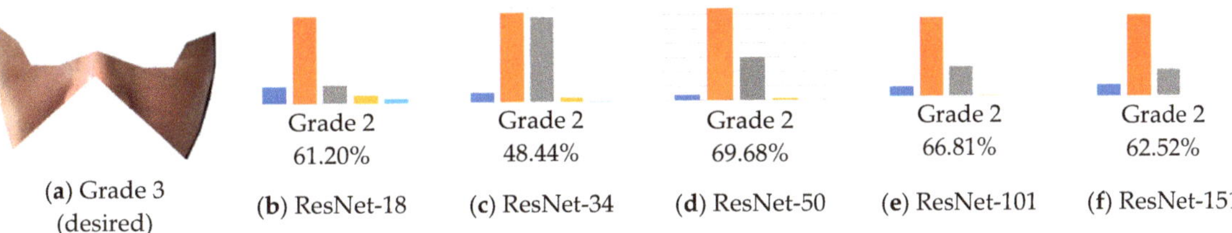

(**a**) Grade 3 (desired)

(**b**) ResNet-18 — Grade 2 61.20%

(**c**) ResNet-34 — Grade 2 48.44%

(**d**) ResNet-50 — Grade 2 69.68%

(**e**) ResNet-101 — Grade 2 66.81%

(**f**) ResNet-151 — Grade 2 62.52%

Figure 10. Example and training results of Grade 3 cheek wrinkles (wrongly classified grades are underlined).

Figure 11 shows Grade 4 cheek speckles. Speckles on the face appear more clearly in the images compared with other skin features. Model classification accuracy is not so good using models with fewer layers. With more layers, model accuracy greatly improves.

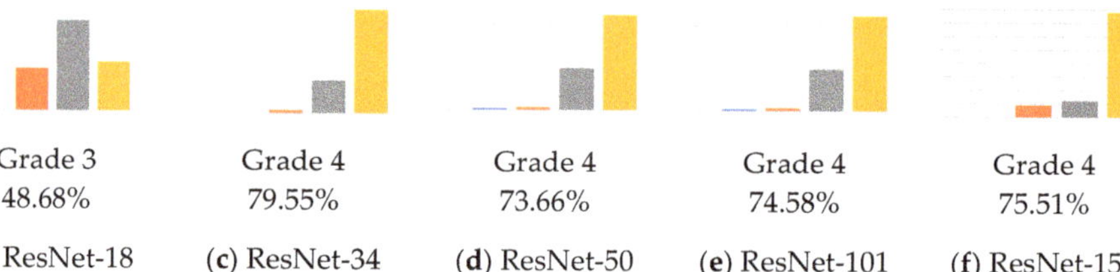

<table>
<tr><td>(a) Grade 4
(desire)</td><td>Grade 3
48.68%
(b) ResNet-18</td><td>Grade 4
79.55%
(c) ResNet-34</td><td>Grade 4
73.66%
(d) ResNet-50</td><td>Grade 4
74.58%
(e) ResNet-101</td><td>Grade 4
75.51%
(f) ResNet-151</td></tr>
</table>

Figure 11. Example and training results of the Grade 4 cheek speckles (wrongly classified grades are underlined).

Figure 12 shows Grade 5 forehead wrinkles. Despite the clarity of these wrinkles, ResNet-18 gives only a fair accuracy in classification. With ResNet-34 or more layers, classification accuracy improves steadily.

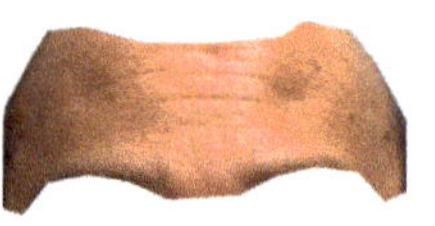

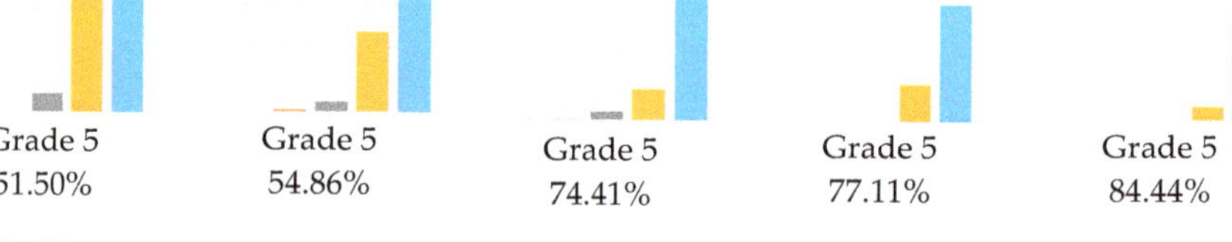

<table>
<tr><td>(a) Grade 5
(desired)</td><td>Grade 5
51.50%
(b) ResNet-18</td><td>Grade 5
54.86%
(c) ResNet-34</td><td>Grade 5
74.41%
(d) ResNet-50</td><td>Grade 5
77.11%
(e) ResNet-101</td><td>Grade 5
84.44%
(f) ResNet-151</td></tr>
</table>

Figure 12. Example and training results of Grade 5 forehead wrinkles.

In brief, as the number of model layers increases progressively, the accuracy increases in parallel. As expected, model performance is closely related to its number of layers (Figure 13).

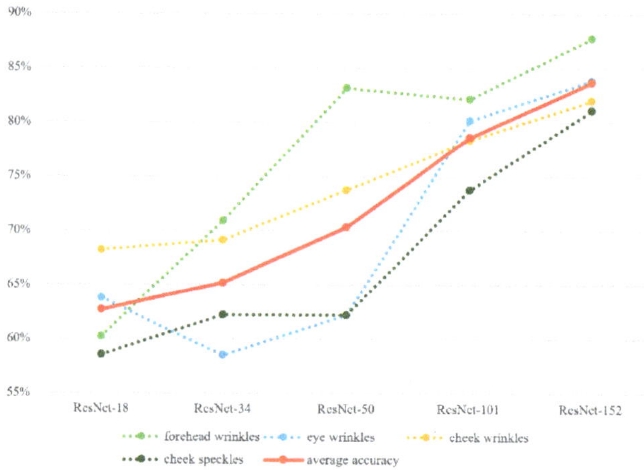

Figure 13. Model accuracy results (from K-fold cross validation) plotted against model layers and facial features.

3.3.2. Results of Model Classification Using Polygon Images Presented in Terms of Precision and Recall

Figure 14 shows results of precision, and Figure 15 shows results of recall. The overall pictures are basically similar to those of accuracy (Figure 13), despite some minor disparities. These disparities are likely related to characteristics of the dataset, such as sample size, hair or shadow noise, and relative abundance of skin features across facial regions. Again, as the number of model layers increases, the precision or recall rate increases.

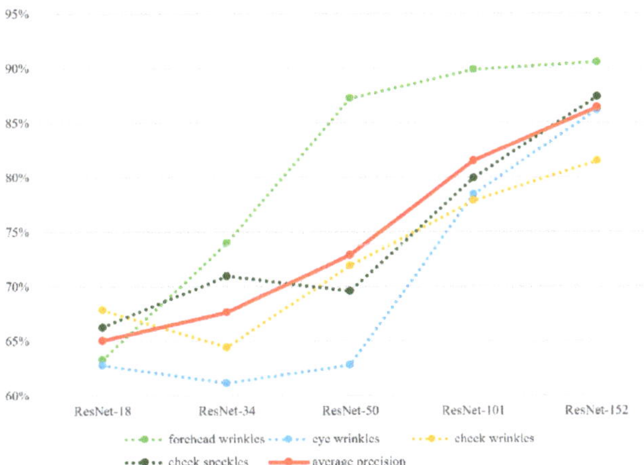

Figure 14. Results of classification precision (from K-fold cross validation) plotted against model layers and facial features.

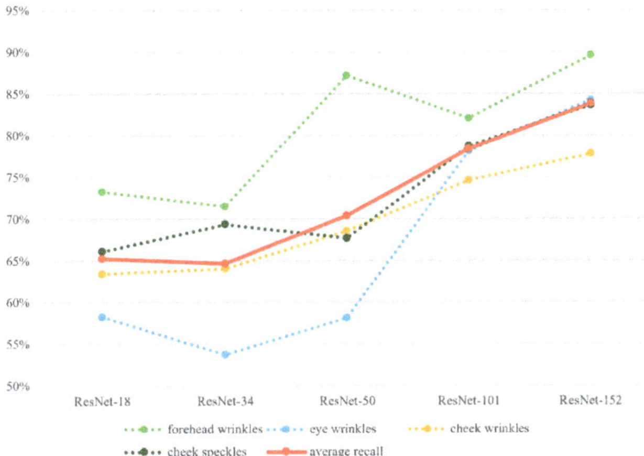

Figure 15. Results of classification recall (from K-fold cross validation) plotted against model layers and facial features.

3.3.3. Training Results Using Whole Face Images

Because the whole face has a larger picture size, training takes more time (at least twice as long) when compared with training using the smaller polygon images. Figure 16 clearly shows that the average accuracy using polygon images is much higher than that using whole face images. Such discrepancy is likely due to interference of the non-feature regions in the whole face images.

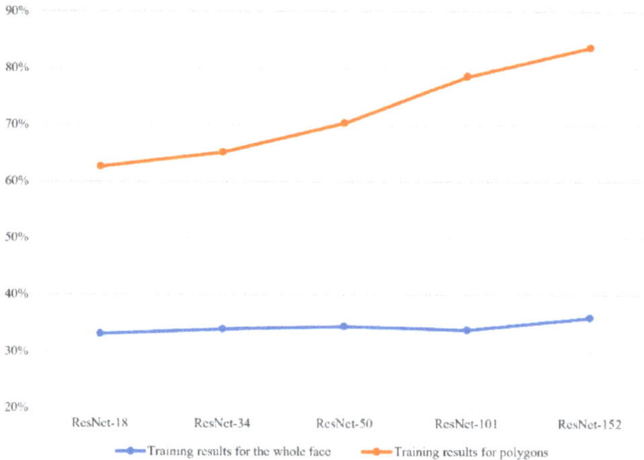

Figure 16. Average accuracy results (from K-fold cross validation), comparing training with the whole face and with polygon images.

3.3.4. Extended Results Using Other Models

After our study with Resnet, other models like AlexNet, and VGG were also tested. Figure 17 confirms that training results using Resnet152 are still the best. These other models give high correct rates comparable to those from Resnet50.

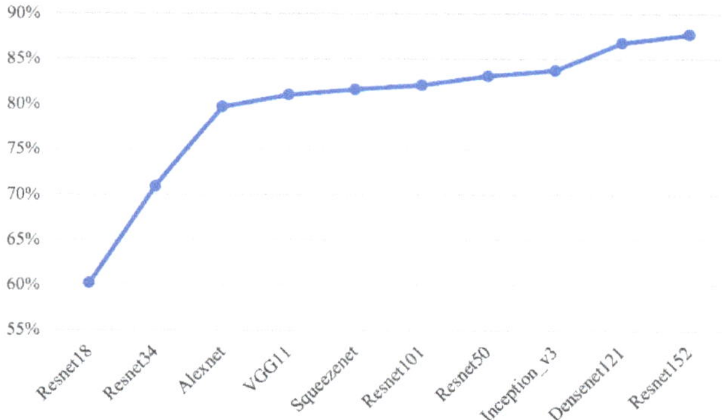

Figure 17. Accuracy results (from K-fold cross validation) on forehead wrinkles plotted across several models.

3.3.5. Limitations

Two main limitations of the study are as follows: (a) the data augmentation step undoubtedly introduced strong bias. If we had larger datasets (e.g., 20,000 original images) our conclusions would definitely be more convincing; (b) we have not compared other models exhaustively. For example, after segmenting the face into various regions, we did not continue with the face detection model for wrinkle and speckle detection and grading classification. The reason for this is technical, as we needed to manually mark and grade the features on the images, a process that is labor-intensive and has questionable accuracy. We therefore do not rule out the possibility of a better performance by other models that we have not fully tested.

4. Conclusions

This is the first deep learning modeling study on classifying grades of two common skin conditions (wrinkles and speckles) using polygon images. The overall test results from classifying such conditions using Resnet models are satisfactory. Deep learning using polygon images produces better results than those using whole face images. A greater number of layers in ResNet produces better performance. ResNet-152 so far shows the best results compared with the other models we tested.

Author Contributions: T.-R.C.: Problem conceptualization, Methodology, Data analysis, Writing-Review and edit final draft, Results tabulate and graphic presentation. M.-Y.T.: Data collection, Software development and execution, Investigation. All authors have read and agreed to the published version of the manuscript.

Funding: This study was supported in part by the A^2IBRC, STUST from the Higher Education Sprout Project of the Ministry of Education, Taiwan, and MOST 109-2221-E-218 -022—from the Ministry of Science and Technology, Taiwan.

Informed Consent Statement: Informed consent was obtained from all participants involved in the study.

Conflicts of Interest: The authors declare no conflict of interest.

References

1. Asakura, K.; Nishiwaki, Y.; Milojevic, A.; Michikawa, T.; Kikuchi, Y.; Nakano, M.; Iwasawa, S.; Hillebrand, G.; Miyamoto, K.; Ono, M.; et al. Lifestyle factors and visible skin aging in a population of Japanese elders. *J. Epidemiol.* **2009**, *19*, 251–259. [CrossRef] [PubMed]
2. 5.1 Layers of the Skin—Anatomy and Physiology 2e | OpenStax. Available online: https://openstax.org/books/anatomy-and-physiology-2e/pages/5-1-layers-of-the-skin (accessed on 2 November 2022).
3. Riaz, F.; Naeem, S.; Nawaz, R.; Coimbra, M. Active contours based segmentation and lesion periphery analysis for characterization of skin lesions in dermoscopy images. *IEEE J. Biomed. Health Inform.* **2019**, *23*, 489–500. [CrossRef] [PubMed]
4. Albert, B.A. Deep learning from limited training data: Novel segmentation and ensemble algorithms applied to automatic melanoma diagnosis. *IEEE Access* **2020**, *8*, 31254–31269. [CrossRef]
5. Liao, H. A Deep Learning Approach to Universal Skin Disease Classification. University of Rochester Department of Computer Science, CSC, 2016. Available online: https://studylib.net/doc/14025144/a-deep-learning-approach-to-universal-skin-disease-classi (accessed on 2 November 2022).
6. Esteva, A.; Kuprel, B.; Novoa, R.A.; Ko, J.; Swetter, S.M.; Blau, H.M.; Thrun, S. Dermatologist-level classification of skin cancer with deep neural networks. *Nature* **2017**, *542*, 115–118. [CrossRef] [PubMed]
7. Lopez, A.R.; Giro-i-Nieto, X.; Burdick, J.; Marques, O. Skin lesion classification from dermoscopic images using deep learning techniques. In Proceedings of the 2017 13th IASTED International Conference on Biomedical Engineering (BioMed), Innsbruck, Austria, 20–21 February 2017; pp. 49–54.
8. Huang, H.; Kharazmi, P.; McLean, D.I.; Lui, H.; Wang, Z.J.; Lee, T.K. Automatic Detection of Translucency Using a Deep Learning Method from Patches of Clinical Basal Cell Carcinoma Images. In Proceedings of the 2018 Asia-Pacific Signal and Information Processing Association Annual Summit and Conference (APSIPA ASC), Honolulu, HI, USA, 12–15 November 2018; pp. 685–688.
9. Anantharaman, R.; Velazquez, M.; Lee, Y. Utilizing Mask R-CNN for Detection and Segmentation of Oral Diseases. In Proceedings of the 2018 IEEE International Conference on Bioinformatics and Biomedicine (BIBM), Madrid, Spain, 6 December 2018; pp. 2197–2204.
10. Mukaida, S.; Ando, H. Extraction and Manipulation of Wrinkles and Spots for Facial Image Synthesis. In Proceedings of the Sixth IEEE International Conference on Automatic Face and Gesture Recognition, Seoul, Korea, 17–19 May 2004; pp. 749–754.
11. Ng, C.; Yap, M.H.; Costen, N.; Li, B. Wrinkle detection using hessian line tracking. *IEEE Access* **2015**, *3*, 1079–1088. [CrossRef]
12. Çanak, B.; Kamaşak, M.E. Automatic Scoring of Wrinkles on the Forehead. In Proceedings of the 2017 25th Signal Processing and Communications Applications Conference (SIU), Antalya, Turkey, 15–18 May 2017; pp. 1–4.
13. Zaghbani, S.; Boujneh, N.; Bouhlel, M.S. Facial Emotion Recognition for Adaptive Interfaces Using Wrinkles and Wavelet Network. In Proceedings of the 2017 IEEE/ACS 14th International Conference on Computer Systems and Applications (AICCSA), Hammamet, Tunisia, 30 October–3 November 2017; pp. 342–349.
14. OpenCV: Cascade Classifier Training. Available online: https://docs.opencv.org/3.4/dc/d88/tutorial_traincascade.html (accessed on 25 August 2021).
15. Peng, C.; Gao, X.; Wang, N.; Li, J. Superpixel-based face sketch–photo synthesis. *IEEE Trans. Circuits Syst. Video Technol.* **2017**, *27*, 288–299. [CrossRef]
16. Nirkin, Y.; Masi, I.; Tuan, A.T.; Hassner, T.; Medioni, G. On Face Segmentation, Face Swapping, and Face Perception. In Proceedings of the 2018 13th IEEE International Conference on Automatic Face Gesture Recognition (FG 2018), Xi'an, China, 15–19 May 2018; pp. 98–105.
17. Facial Landmarks with Dlib, OpenCV, and Python—PyImageSearch. Available online: https://pyimagesearch.com/2017/04/03/facial-landmarks-dlib-opencv-python/ (accessed on 2 November 2022).

18. Khan, K.; Mauro, M.; Migliorati, P.; Leonardi, R. Head Pose Estimation through Multi-Class Face Segmentation. In Proceedings of the 2017 IEEE International Conference on Multimedia and Expo (ICME), Hong Kong, China, 10–14 July 2017; pp. 175–180.
19. Benini, S.; Khan, K.; Leonardi, R.; Mauro, M.; Migliorati, P. Face analysis through semantic face segmentation. *Signal Process. Image Commun.* **2019**, *74*, 21–31. [CrossRef]
20. Hassan, S.Z.; Ahmad, K.; Al-Fuqaha, A.; Conci, N. Sentiment Analysis from Images of Natural Disasters. In *Image Analysis and Processing—ICIAP 2019*; Ricci, E., Rota Bulò, S., Snoek, C., Lanz, O., Messelodi, S., Sebe, N., Eds.; Springer International Publishing: Cham, Switzerland, 2019; pp. 104–113.
21. Niko 81 Facial Landmarks Shape Predictor | Github. Available online: https://github.com/codeniko/shape_predictor_81_face_landmarks (accessed on 2 November 2022).
22. He, K.; Zhang, X.; Ren, S.; Sun, J. Deep residual learning for image recognition. *arXiv* **2015**, arXiv:1512.03385.

Article

Repetition with Learning Approaches in Massive Machine Type Communications

Li-Sheng Chen [1,*], Chih-Hsiang Ho [2], Cheng-Chang Chen [3], Yu-Shan Liang [4] and Sy-Yen Kuo [5]

[1] Department of Communications Engineering, Feng Chia University, Taichung 407, Taiwan
[2] Institute for Information Industry, Taipei 106, Taiwan
[3] Bureau of Standards, Metrology and Inspection, M. O. E. A., Taipei 100, Taiwan
[4] Department of Electronic Engineering, National Kaohsiung University of Science and Technology, Kaohsiung 824, Taiwan
[5] Department of Electrical Engineering, National Taiwan University, Taipei 100, Taiwan
* Correspondence: lschen@fcu.edu.tw

Abstract: In the 5G massive machine type communication (mMTC) scenario, user equipment with poor signal quality requires numerous repetitions to compensate for the additional signal attenuation. However, an excessive number of repetitions consumes additional wireless resources, decreasing the transmission rate, and increasing the energy consumption. An insufficient number of repetitions prevents the successful deciphering of the data by the receivers, leading to a high bit error rate. The present study developed adaptive repetition approaches with the k-nearest neighbor (KNN) and support vector machine (SVM) to substantially increase network transmission efficacy for the enhanced machine type communication (eMTC) system in the 5G mMTC scenario. The simulation results showed that the proposed repetition with the learning approach effectively improved the probability of successful transmission, the resource utilization, the average number of repetitions, and the average energy consumption. It is therefore more suitable for the eMTC system in the mMTC scenario than the common lookup table.

Keywords: massive machine type communications (mMTC); enhanced machine type communication (eMTC); repetition; machine learning; k-nearest neighbor (KNN); support vector machine (SVM)

Citation: Chen, L.-S.; Ho, C.-H.; Chen, C.-C.; Liang, Y.-S.; Kuo, S.-Y. Repetition with Learning Approaches in Massive Machine Type Communications. *Electronics* **2022**, *11*, 3649. https://doi.org/10.3390/electronics11223649

Academic Editor: Christos J. Bouras

Received: 1 September 2022
Accepted: 19 October 2022
Published: 8 November 2022

Publisher's Note: MDPI stays neutral with regard to jurisdictional claims in published maps and institutional affiliations.

1. Introduction

Machine type communication (MTC) describes data exchange and communications between machines and is a crucial component in areas such as smart cities, traffic optimization, smart poles, e-medicine, and the industrial Internet of Things (IoT) [1,2]. Following the increase in MTC applications, the number of connected wireless devices has risen dramatically; the number of wireless MTC devices is expected to reach several billion [3]. However, this has created a critical challenge for the devices' network-access capability. Massive MTC (mMTC) is the primary third-generation partnership project (3GPP) fifth-generation (5G) application scenario and technology implemented in IoT devices [4]. The respective technology involves connecting a massive number of components. For each square kilometer of area in a developed region, the machine–machine communications between 1 million devices are expected. Such communications are primarily for small quantities of data with relatively high latency tolerance. The components must be cost-efficient and equipped with long battery lives.

3GPP enhanced MTC (eMTC) and 3GPP narrowband IoT (NB-IoT) are 5G IoT technology standards that address the requirements of 3GPP 5G mMTC application scenarios. Studies have explored the physical characteristics and design goals of eMTC and NB-IoT [5–7]. In the long-term evolution (LTE), the 5G, advanced eMTC, and NB-IoT user equipment (UE) switches from idle mode to connected mode through random access (RA) [8,9]; this process is achieved through a four-way handshake, which involves a preamble transmission, an RA response, a connection request, and a connection resolution. When the UE transmits data in a highly synchronized manner, the signaling capacity of the evolved node B can be easily exceeded, causing severe system congestion [10]. Ali and Hamouda [11] proposed an algorithm involving beehive search and initial synchronization. The simulation results confirmed that the algorithm provides satisfactory network efficacy even for devices with an extremely low signal-to-noise ratio. In [12], a maximum-likelihood detector was applied for initial timed collection in IoT devices. The average detection latency of the detector was half as high as that of the related method, and the energy required for each timed collection decreased by 34%. An approach integrating agent-based modeling and simulation was employed for IoT system efficacy analysis in [13,14]. The agent-based cooperative smart object [15] approach uses OMNeT++ for simulation. Liu et al. [16,17] proposed two novel resource coordination methods that involve bridging dynamic sensing tasks with heterogeneous IoT sensors and controlling the operating cycles of devices to save energy. In [18], eMTC infrastructure coverage efficacy was analyzed and compared with that of LTE technology to reinforce the eMTC coverage at 15 dB.

The design of mMTC technology must fulfill the following four requirements:

a. Coverage: The coverage of mMTC must attain a maximum coupling loss (MCL) of 164 dB [19]. Even when the signal from the transmitter to the receiver is attenuated by as much as 164 dB, the receiver must successfully decipher the packet. In addition, because increasing coverage through repeat transmission reduces the data transmission rate substantially, 5G mMTC coverage must also be successfully implemented at a transmission rate as high as 160 bit/s. Therefore, selecting an appropriate coverage enhancement (CE) level and number of repetitions for the signal quality is critical.

b. UE battery life: Devices with long battery lives are required in 5G MTC applications such as smart electric and water meters. These devices may be installed in environments that impede battery replacement or in which the cost of a battery replacement is excessive. Therefore, mMTC devices must have battery lives of no less than 10 years [20], similar to those of eMTC or NB-IoT devices.

c. Connection density: As the demand for IoT applications increases, the density of 5G IoT devices is expected to reach approximately 1 million devices per km^2 in developed areas. Therefore, 5G mMTC must support a connection density value as high as 106 devices per km^2 while maintaining a specific quality of service.

d. Latency: Although most MTC devices have considerable data transmission latency tolerance, the 5G mMTC specifications include a latency tolerance requirement to ensure satisfactory quality of service. Specifically, for each 20-byte application layer packet transmitted by a device, the latency should not exceed 10 s in a channel with an MCL of 164 dB.

Both NB-IoT and eMTC are low-power wide-area network technologies in authorized spectra. NB-IoT is versatile in spectra and can support three modes of deployment. However, eMTC is faster and exhibits a broader range of applications; in a half-duplex system, the uplink and downlink speed of eMTC is 375 kbps, making eMTC applicable for IoT applications for which a medium data rate is required. Regarding peak speed, NB-IoT exhibits almost no mobility because it does not support handover between base stations, whereas eMTC exhibits more favorable mobility.

Several papers have proposed the study and overview of machine learning technologies, applications, and challenges for IoT and 5G networks [21,22]. Ref. [23] proposes the method with a game and transport theoretic approach for a fog load balancing problem. The work provides a feasible and efficient load balancing solution to ensure an optimal job assignment in the fog computing network with the NB-IoT. One study [24] used machine learning techniques that can be applied for the automation of network functions in 5G network slicing. The intelligent station recognition scheme with support vector machines (SVMs) has been proposed to achieve the fine management of stations [25]. A scheduler framework using reinforcement learning has been proposed [26]; the appropriate scheduling strategy is able to maximize user satisfaction, measured in terms of the distinct quality of the service requirements. Ref. [27] proposes the learning approach to implicitly extract channel features and recover tag symbols using the deep transfer learning (DTL) approach. In [28], a spectrum management architecture was studied and a machine learning–based spectrum decision framework for the IoT network was proposed.

The eMTC requirements support two CE levels (CE Mode A and CE Mode B) [29] and several numbers of repetition, and each piece of UE can select the appropriate CE levels and number of repetitions according to the signal quality. UE with inferior signal quality requires numerous repetitions to compensate for additional signal attenuation, as depicted in Figure 1, which shows the CE level and repetitions for eMTC.

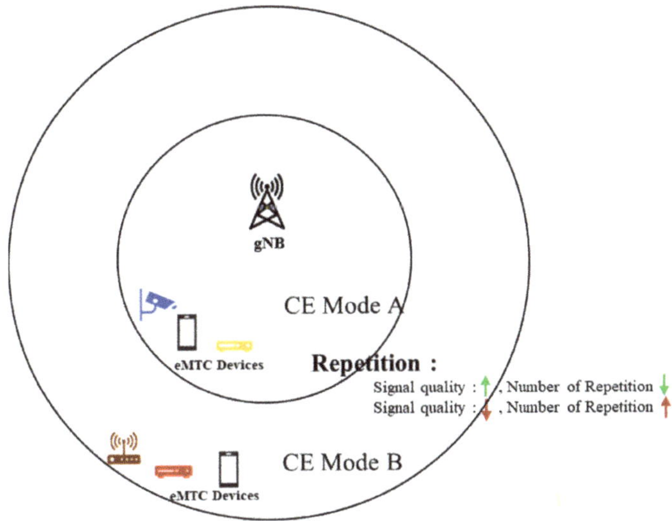

Figure 1. CE level and repetitions for eMTC.

In the eMTC system and mMTC application scenario, due to the main consideration of coverage and power consumption, UE pieces with poor signal quality will use more repetitions to compensate for additional signal attenuation. For the selection of the UE's CE level and repetition times, if they are too high they will waste valuable wireless resources, reduce the transmission rate, and consume more power; if they are too low, the data may not be successfully resolved at the receiving end, resulting in a higher BER (bit error rate). Therefore, we propose the learning adaptive repetition approaches based on k-nearest neighbor (KNN) and the support vector machine (SVM) for an eMTC system in an mMTC application scenario. KNN is easy to implement, has high accuracy, and is insensitive to outliers. When KNN selects appropriate training parameters, the KNN approach can achieve high discrimination accuracy. SVM can avoid the neural network structure selection and local minima problem. SVM does not have a general solution to nonlinear problems; therefore, we significantly choose the kernel function to handle it. For the selection of the CE level and repetition times, we adopted KNN and SVM to propose a learning-based repetition approach, which effectively saves energy efficiency and improves overall network transmission performance. In the present study, a repetition with the learning approach (RLA) was proposed and developed with the aim of substantially increasing the network transmission efficacy. The main contributions of this paper are summarized as follows: (1) we study the problem of repetition number selection for the eMTC system in the mMTC scenario, where the main goal is to minimize the average energy consumption and block error rate (BLER) and maximize the successful transmission probabilities of UE pieces and resource utilization. (2) Due to the limited UE energy efficiency and slave resources of eMTC, we provided the easy-to-implement and learning-based policy for repetition number selection to ensure that eMTC saves energy efficiency and improves transmission efficiency. (3) We adopted the KNN approach and selected the appropriate training parameters meaningfully so that the KNN approach could be easily implemented and could have high accuracy. (4) We adopted the SVM approach to avoid the neural network structure selection and local minima problem and chose the kernel function significantly. (5) Finally, we conducted an extensive numerical analysis to evaluate the performance of the proposed repetition number selection approaches, and the simulation results show that our proposed approaches (RLA–KNN and RLA–SVM) are more efficient than the common lookup table (LUT) method. The proposed repetition with the learning approach can effectively improve the probability of successful transmission, resource utilization, average number of repetitions, and average energy consumption. The remainder of this paper is organized as follows. Section 2 describes the eMTC system. The proposed approaches are presented in Section 3. In Section 4, we evaluate the performance of our proposed approaches through simulation. Finally, Section 5 provides the conclusions.

2. eMTC System

eMTC technology involves enhancing and customizing mMTC according to the application requirements, such as repetition, scheduling, discontinuous reception, control channels, and system information block (SIB). The eMTC UE monitors the same master information blocks (MIBs), the scheduling period of which is 40 ms. With eMTC, however, within 40 ms, the MIB information is transmitted at the 0th, 18th, and 19th time slots in each radio frame. As such, the coverage of a physical broadcast channel is enhanced through repetition.

In the 3GPP protocol, eMTC independently defines the SIB information; that is, SIB–BR is composed of SIB1–BR and other numbers of SIB information. SIB–BR features a time–frequency position different from that of LTE. SIB1–BR is transmitted on a physical downlink shared channel (PDSCH) at a fixed period of eight radio frames (80 ms). Within each period, SIB1–BR is repeatedly transmitted 4, 8, or 16 times according to the designation by an MIB.

SIB1–BR is transmitted without a controlled schedule through frequency hopping between NBs. The pattern of the frequency hopping is related to the system bandwidths and physical-layer cell identities. The other SIB–BR information is transmitted in the PDSCH; the system information is transmitted without controlled schedules. The schedule information of the system information, such as the time–frequency locations, modulation coding scheme levels, and number of repetitions, is determined in SIB1–BR.

When changes occur in the broadcast information, except for SIB10–BR, SIB11–BR, SIB12–BR, and SIB14–BR, an update is announced through paging or DCI6-2 during the period of change, and new broadcast information is transmitted in the subsequent period of change. The same procedure applies when changes occur in SIB10–BR, SIB11–BR, SIB12–BR, and SIB14–BR.

Unlike NB-IoT, eMTC supports the switching of connection statuses and uninterrupted service transmission when the UE moves across regions. Therefore, eMTC is applicable for services such as voice calls and logistics tracking. The energy-saving function of eMTC is achieved through extension discontinuous reception, power saving mode, and the reduction in the periodic position update frequency. The 3GPP system extends the time of the periodic position update procedure to reduce the periodic position update frequency of the UE, thereby mitigating the signal load in the networks and the power expense in the UE. With the power saving mode of the 3GPP R12 system, the UE timer enters a deep sleep mode after the UE's tracking area is updated and its attach procedure is completed. In the deep sleep mode, the UE does not monitor paging, and the radio transceiver unit is switched off to save significantly more power than that saved in the idle mode. The UE remains registered in the network in the deep sleep mode without the requirement of rerunning the attach procedure or reestablishing the packet data network connection.

The physical layer of eMTC is redesigned through 3GPP, and the eMTC signal coverage is enhanced through the repetition mechanism in the physical channel. In downlink, eMTC employs neither a physical control format indicator channel nor a physical hybrid automatic repeat request indicator channel. The number of signals transmitted in the 40 ms period is increased in the physical broadcast channel. Rather than the conventional LTE physical downlink control channel, an MTC physical downlink control channel is employed for a maximal repetition number of 256. A maximum of six RBs can be applied by a UE unit in a PDSCH. The modulation coding scheme level and the maximal number of repetitions are restricted; specifically, the maximal number of repetitions is set as 2048. Similarly, a maximum of six RBs can be applied by a UE unit in a physical uplink shared channel (PUSCH). The modulation coding scheme level and the maximal number of repetitions are restricted. In particular, the maximal number of repetitions is set as 2048. In a physical random access channel, a maximum of six RBs can be applied by a UE unit, and a maximum of 128 repetitions can be performed. In a physical uplink control channel (PUCCH), a maximum of 32 repetitions can be performed. The coverage modes of the UE connected to a network can be categorized into two types, namely CE mode A, which indicates satisfactory coverage with no or few repetitions, and CE mode B, which involves more repetitions. In CE mode A, the maximal number of repetitions is 32 in a PDSCH or a PUSCH and 8 in a PUCCH; in CE mode B, the maximal number of repetitions is 2048 in a PDSCH or a PUSCH and 32 in a PUCCH.

3. Repetition with Learning Approaches

Most repetition approaches involve determining the number of repetitions according to the one-dimensional channel quality measured in a data table generated through a simulation. This method is strongly reliant on specific channel models, and errors in the channel quality measurement can lead to the selection of an incorrect number of repetitions. Moreover, a large-dimension data table requires considerable memory space. With the future trend of repetition technology development, one-dimensional channel quality indicators can no longer comprehensively represent the conditions of the channels in complex systems. This is because the link efficacy is related to numerous system parameters, and this can lead to erroneous channel quality assessment and inhibited system efficiency. The accuracy of channel quality indicators can be improved through an increase in the dimension of the channel quality. Machine learning provides a system with high-dimensional channel quality indicators that take into account changes in the environment. The relationship between channel quality and link reliability can be identified through examining the historical records in data transmission, thereby attaining an accurate prediction of the number of repetitions required according to the channel quality measurement results. Conventional repetition approaches are based on inquired data tables and cannot be updated according to system environments; this reduces their flexibility. Machine learning enables a system to learn and adapt to changes in an environment and flexibly adjust its channel quality selection standards, thereby accurately predicting the number of repetitions required. Accordingly, research on the implementation of machine learning to predict the number of uplink repetitions in eMTC is paramount for system efficacy enhancement.

K-nearest neighbor (KNN) is a supervised learning approach in which a particular sample is considered to belong to a specific class if most of the k-nearest neighbors to it in its eigenspace also belong to said class [30–32]. In particular, a non-classified sample is categorized to the class in which its nearest classified neighbors belong. When new samples enter a group of classified samples, their distances from the training data are calculated, and the k-nearest samples are determined. Subsequently, the class with the greatest number of neighbors among the k-selected data is determined, and the new sample is classified accordingly.

In the training phase, according to the feature set acquired from the signal-to-interference-plus-noise ratio (SINR) of each subcarrier, number i, which corresponds to repetition number, is selected to optimize the data rate and categorize the training set. All of the elements in each feature set must satisfy the optimized class, which maximizes the number of data transmitted, subject to the limitations of the block error rate.

$$\arg \max_{i}\{RN_i : BLER_i < H\} \tag{1}$$

RLA-KNN approach training requires a training set with SINR feature vectors. Each vector is assigned to a number i according to equation (1). In RLA-KNN, when the training sample set and the K value and distance are determined, the class of any new sample can be determined. $U = \{u_1, u_2, \ldots, u_n\}$, representing the eigenvector training set, and S = $\{s_1, s_2, \ldots, s_n\}$, representing the class to which each eigenvector corresponds. RLA-KNN approach training is performed to identify all the vectors in a feature set and the approximated repetition approach corresponding to them.

Let the training sample set TSS in KNN be

$$\text{TSS} = \{(s_1,\, u_1),\, (s_2,\, u_2), \ldots, (s_n,\, u_n)\} \tag{2}$$

The sample to be classified, x, is imported. The distance between x and all other samples in the training set can be calculated as follows: $L(x,\, s_j)$, $j = 1, 2, \ldots, n$. According to the distance data, K samples in training set that are closest to x are determined and referred to as $Near_k(x)$.

Figure 2 illustrates the RLA using KNN (RLA–KNN), where k is the number of comparisons, and $L(x,\, s_j)$ is a measured distance.

RLA–KNN

Input: U, S and X

Output: RN

Step 1. Let the training sample set TSS in KNN be the equation (2).

Step 2. Calculate the distance between a point and a point of a known class $L(x, s_j)$, where $j = 1, 2, \ldots, n$ by equation (3) where $p = 2$.

Step 3. Sort in ascending order of distance

Step 4. Select the K points with the smallest distance from the points to be classified, and determine the number of occurrences of the class in which the last K points.

Step 5. Return the class with the highest number of occurrences of the last K points as the predicted classification of the points to be classified.

Step 6. Select the parameters corresponding to the predicted classification (repetition number) to set for transmission.

Figure 2. RLA–KNN approach.

In the KNN approach, distance measurement is the most effective approach to determining the similarity between two samples. Commonly used distance measures include the Euclidean distance, Manhattan distance, Chebyshev distance, and Minkowski distance.

Let the eigenspace X be an n-dimension real vector space (R^n): $x_i, x_j \in X$; $x_i = \left(x_i^1, x_i^2, \ldots, x_i^n\right)^T$; and $x_j = \left(x_j^1, x_j^2, \ldots, x_j^n\right)^T$. The distance between x_i and x_j is defined as [33]:

$$L_p\left(x_i,\, x_j\right) = \left(\sum_{l=1}^{n}\left|x_i^{(l)} - x_j^{(l)}\right|^p\right)^{1/p} \tag{3}$$

When $p \geq 1$, the Minkowski distance is derived. When $p = 1$, the Manhattan distance is derived. When $p = 2$, the Euclidean distance is obtained. When $p = \infty$, the Chebyshev distance is derived. In the RLA-KNN, we set the p to 2, using distance measures by Euclidean distance.

The time complexity of the prediction for the RLA–KNN approach is O(N), and the time complexity of sorting for the RLA–KNN approach is O(N log N). KNN approaches are generally inapplicable for large-scale communication systems because of the complex calculation procedure involved. Therefore, SVMs are implemented to identify the repetition approach according to high-dimensional channel quality measurement results. SVM is a commonly employed machine learning method, which is generally modeled as a convex quadratic programming problem [34,35]. Therefore, an SVM is simpler to execute than the KNN approach.

Figure 3 depicts the RLA using an SVM (RLA–SVM) approach, in which a repetition approach is selected according to the BLER in the data.

RLA–SVM

Input: X, Y

Output: RN

Step 1. Let the training sample set in SVM be the equation (4).

Step 2. The new dataset generated by the equation (5) is linearly separable .

Step 3. Choose a function: $k(x_1, x_2) = \langle x_1, x_2\rangle^2$, use it to replace the inner product calculation of SVM.

Step 4. For the new dataset after dimension expansion, use equation (6) and equation (7) to predict prediction y.

Step 5. Select the parameters corresponding to the predicted classification (repetition number) to set for transmission.

Figure 3. RLA–SVM approach.

For a set of data X and label Y, the task of SVM is to find a set of parameters such that $x^{\theta^T} = threshold$, and the samples with $x^{\theta^T} < threshold$ are judged as negative samples, and the samples with $x^{\theta^T} > threshold$ are judged as positive samples. The sample X is a two-dimensional data set of channel quality:

$$X = \begin{bmatrix} x_1^{(1)} & x_2^{(1)} \\ x_1^{(2)} & x_2^{(2)} \\ \vdots & \vdots \\ x_1^{(m)} & x_2^{(m)} \end{bmatrix} \tag{4}$$

This dataset X is linearly inseparable on the two-dimensional plane, and the new dataset generated by a transformation function $\phi(x)$ is linearly separable:

$$\phi(x_1, x_2) = \left(x_1^2, \sqrt{2}x_1, x_2, x_2^2 \right) \tag{5}$$

For the new dataset after the dimension increase, the prediction y for repetition number can be expressed as:

$$\begin{aligned}
y_{predition} &= \sum_{i=1}^{m} \lambda_i y^{(i)} \langle \phi(\hat{x}),\ \phi(x^{(i)}) \rangle \\
&= \sum_{i=1}^{m} \lambda_i y^{(i)} \langle ((\hat{x}_1^2, \sqrt{2}_1, \hat{x}_2, \hat{x}_2^2)),\ (x_1^{(i)^2}, \sqrt{2}x_1^{(i)}, x_2^{(i)}, x_2^{(i)^2}) \rangle \\
&= \sum_{i=1}^{m} \lambda_i y^{(i)} (\hat{x}_1^2, x_1^{(i)^2} + 2\hat{x}_1 x_1^{(i)} \hat{x}_2 x_2^{(i)} + \hat{x}_2^2, x_2^{(i)^2}) \\
&= \sum_{i=1}^{m} \lambda_i y^{(i)} \langle (\hat{x}_1, \hat{x}_2),\ (x_1^{(1)}, x_2^{(2)}) \rangle^2 \\
&= \sum_{i=1}^{m} \lambda_i y^{(i)} \langle \hat{x}_1, x_1^{(i)} \rangle^2
\end{aligned} \tag{6}$$

Through the above transformation, after the data set is upscaling, the calculation of SVM training and prediction can actually be converted into the calculation of the original feature. We choose a function: $k(x_1, x_2) = \langle x_1, x_2 \rangle^2$, use it to replace the inner product calculation of SVM:

$$\hat{y} = \begin{cases} +1, \ \sum_{i=1}^{m} \lambda_i y^i k(\hat{x}, x^i) \geq +1 \\ -1, \ \sum_{i=1}^{m} \lambda_i y^i k(\hat{x}_1, x^i) \leq +1 \end{cases} \tag{7}$$

The training and prediction process of this SVM model is equivalent to doing it in a high-dimensional space, which achieves the purpose of linearly dividing the data set, and does not add complex operations, among which is the kernel function.

The loss function of SVM is:

$$\begin{aligned}
Loss &= \tfrac{1}{2}\|\theta_2\| + \sum_i max(0, 1 - y\hat{y}) \\
&= \tfrac{1}{2}\|\theta_2\| + \sum_i max(0, 1 - y(x\theta + \theta_0))
\end{aligned} \tag{8}$$

The predicted result corresponds to a repetition number. Select the parameters corresponding to the predicted classification (repetition number) to set for transmission.

4. Performance Evaluation

The RLA was applied to evaluate efficacy. The performances for RLA–KNN and RLA–SVM were analyzed subject to various parameters for a single piece of UE, evaluated in a simulated environment and compared with those of the common lookup tables (LUTs) from the eMTC system optimization. This analysis included the successful transmission probabilities of the UE, the average numbers of the repetitions, the resource utilization, and the average energy consumption.

The traffic model defined in 3GPP TR 36.763 [36] was used as the traffic model in the simulation, where the number of eMTC devices was set to 20,000–200,000 per sector; the average number of uplink traffic reports was set to 20/s; and the payload size of each device ranged from 20 to 200 bytes. In the simulation, each point represented the average of 10,000 samples, each of which was acquired between the first uplink transmission and the end of the last uplink transmission (i.e., the observation interval). The relevant simulation environment parameters are illustrated in Table 1. The indicators used to evaluate the performance of the proposed approach were as follows: (1) BLER versus signal to interference plus noise ratio (SINR) for various k values in RLA–KNN using a single piece of UE; (2) throughput versus SINR for the various k values in RLA–KNN using a single piece of UE; (3) BLER versus SINR with the RLA–SVM–radial basis function (RBF), RLA–SVM–linear, and RLA–KNN for a single piece of UE; (4) throughput versus SINR with RLA–SVM–RBF, RLA–SVM–linear, and RLA–KNN for a single piece of UE; (5) the successful transmission probabilities of the UE pieces; (6) the average number of repetitions; (7) the resource utilization; and (8) the average energy consumption.

Table 1. Simulation Parameters.

Parameter	Assumption
Sector per cell	3
Fractional frequency reuse	Hard
Allocation size	12 tones @ 15 KHz
Resource Unit (RU)	1 ms
eMTC Bandwidth	6 PRBs (1.08 MHz)
Channel Model	Typical Urban (TU)
Payload size	20~200
Number of eMTC UE pieces per sector	20,000~200,000
DL Antenna Configuration	gNB: 2Tx/4Tx, UE: 1Rx
Uplink Antenna Configuration	gNB: 2Rx/4Rx, UE: 1Tx
Doppler Spread	1 Hz
gNB Tx power	43 dBm
UE Tx power	20 dBm

The RLA–KNN repetition approach consists of a training phase and a test phase. The training phase requires a training set, each of which corresponds to a unique SINR sequence and is assigned to a repetition number. During the training phase, all the vectors in the training set must undergo all repetitions for accurate classification of the set to the repetition number. The vectors are then classified according to the BLER value of each repetition.

As shown in Figures 4 and 5, a higher k leads to higher system efficacy. However, after k exceeds a certain value, the system efficacy begins to drop. Ordinarily, a higher k limits the effect of erroneous classification on the training results more effectively, yielding a smaller classification error during the test phase and higher system efficacy. However, because of a limit in the size of the training set, the error rate starts to increase after k exceeds a certain value, lowering system efficacy. As shown in the simulation results, the system efficacy was maximized when k was 30; so, this value was employed in the follow-up simulation.

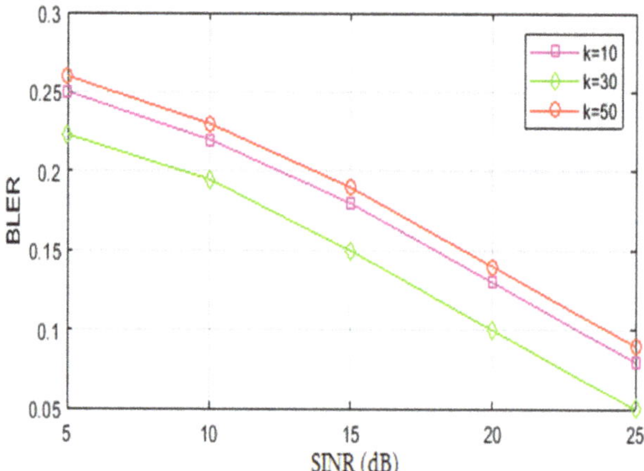

Figure 4. BLER versus SINR for various k values in RLA–KNN.

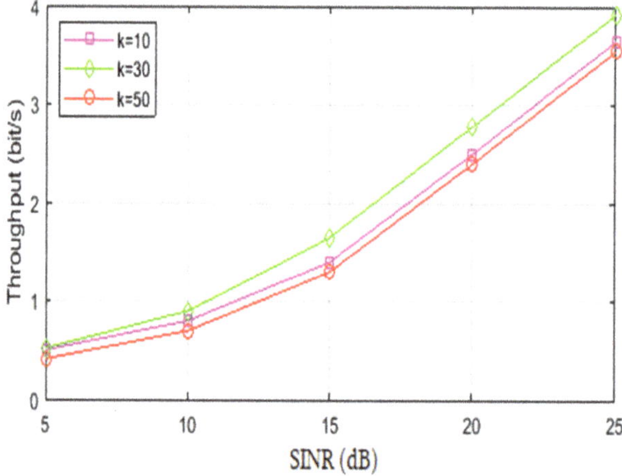

Figure 5. Throughput versus SINR for the various k values in RLA–KNN.

Accordingly, when the number of data in the system is high, the KNN approach requires a complicated calculation procedure and cannot be unsuitably applied in actual network communications without dimensionality reduction. Therefore, the SVM approach is relatively flexible because it can employ numerous kernel functions, reducing the complexity of the calculation procedure considerably. In an RLA–SVM repetition approach, to classify a new sample (the SINR set of a subcarrier), the interference function $h_m(x)$ of each repetition and the BLER of approximately $r_m(h_m(x)$ corresponding to each of the said functions must be identified. We compared the efficacy of the RLA–SVM approach with that of the RLA–KNN approach.

As depicted in Figures 6 and 7, for RLA–SVM the linear kernel function is nearly as efficient as the Gaussian RBF kernel. In particular, the RBF kernel function is slightly more efficient than the linear kernel function, but the RBF kernel function involves a more complicated calculation procedure and requires a larger space for data storage. Moreover, the RLA–SVM repetition approach involves considerably lower calculation and time complexity than does the RLA–KNN repetition approach, even though the two are almost equally efficient in data transmission. Therefore, the RLA–SVM repetition approach is more applicable for an actual eMTC system than the RLA–KNN repetition approach.

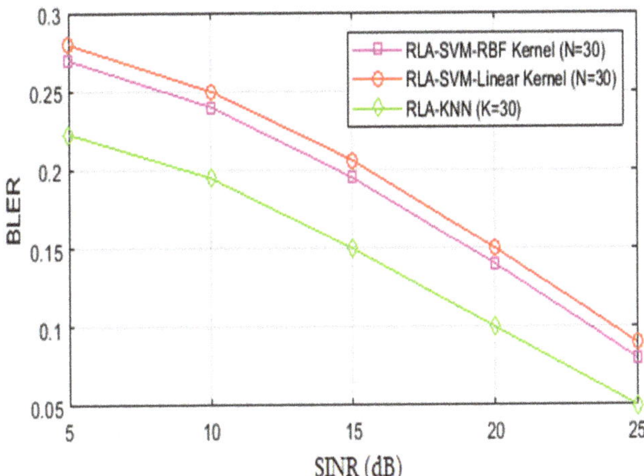

Figure 6. BLER versus SINR for RLA–SVM–RBF, RLA–SVM–linear, and RLA–KNN.

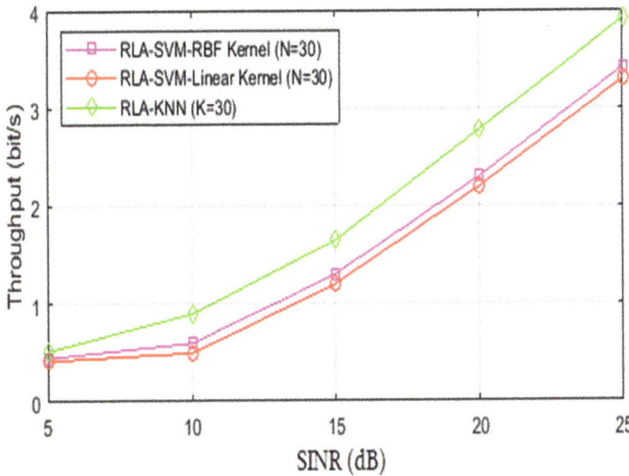

Figure 7. Throughput versus SINR for RLA–SVM–RBF, RLA–SVM–linear, and RLA–KNN.

After analysis of the performance of RLA–KNN and RLA–SVM subject to various parameters for a single piece of UE, to optimize the eMTC system we compare the performances of RLA–KNN, RLA–SVM, and LUT. This analysis includes the probability of successful transmission, the average number of repetitions, resource utilization, and average energy consumption.

The successful transmission probabilities for RLA–KNN, RLA–SVM, and the LUT are shown in Figure 8. The probability of successful transmission gradually decreases as the intensity of the number of pieces of UE per sector increases (i.e., as more users attempt access). Failure probability is observed for high intensity with the use of the LUT method. Apart from collision, the increased failure rate is due to the interference and radio channel effects, caused by the presence of many pieces of UE with wide coverage close to the eNB, which affects such UE pieces farther away in the same coverage zone. The probabilities of successful transmission for RLA–KNN and RLA–SVM are higher than for the LUT method because the repetition number selection policy is considered to be an interference factor from the training data.

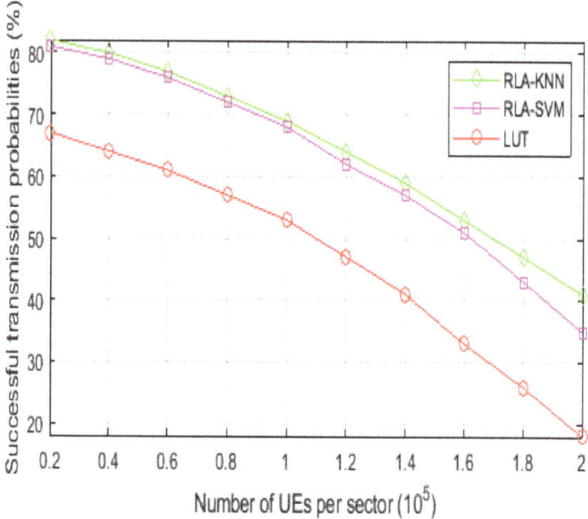

Figure 8. Probability of successful transmission according to UE numbers.

Figure 9 presents the average numbers of repetitions in the RLA–KNN, RLA–SVM, and LUT methods. With the RLA–KNN and RLA–SVM approaches, a significant reduction in repetition is observed. The repetition for the LUT method is noted to be greater than for the RLA–KNN and RLA–SVM methods. In many cases, the LUT method uses excessive numbers of repetitions to achieve successful transmission. In RLA–KNN and RLA–SVM, the number of repetitions required to achieve successful transmission is significantly reduced by the training data and learning processes.

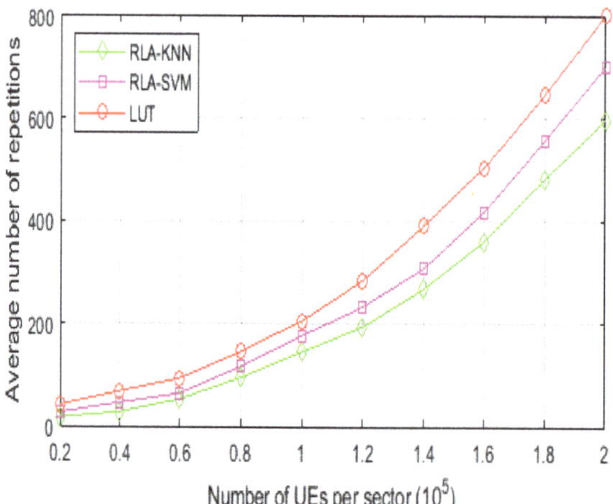

Figure 9. Average number of repetitions.

The resource utilization for RLA–KNN, RLA–SVM, and the LUT is presented in Figure 10. The resource utilization for the three methods is seen to decrease as the number of UE pieces per sector increases. RLA–KNN and RLA–SVM also exhibit higher resource utilization than the LUT because more resources are utilized in the same time period if a shorter repetition is adopted. The result indicates that a smaller number of repetitions is associated with greater resource utilization.

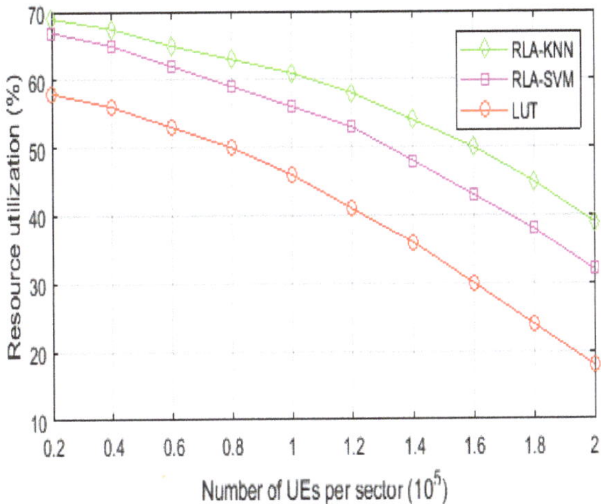

Figure 10. Resource utilization.

Figure 11 presents the average energy consumption per sector in the transmission mode by all users. This decreases the number of repetitions and, thus, power consumption. With the suitable reduction in repetition using the RLA–KNN and RLA–SVM methods, energy consumption is also significantly reduced. A significant reduction in energy consumption is achieved for the uplink compared with for the LUT method. More frequent transmission increases power consumption quickly, which is impractical for IoT devices. More efforts are required to reduce the power consumption of the radio-frequency modules, such as the number of repetitions, under the satisfaction of successful transmission.

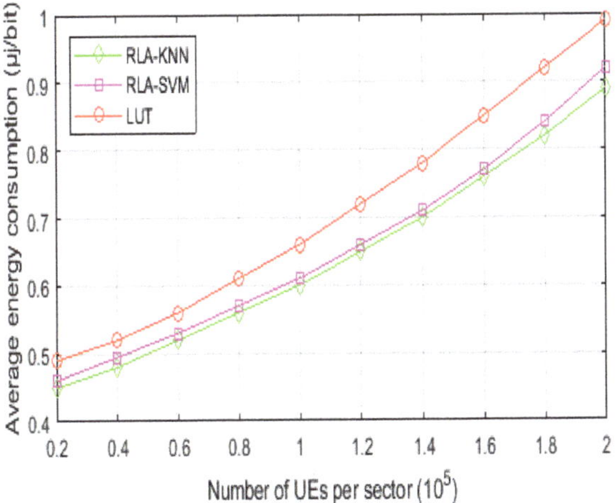

Figure 11. Average energy consumption.

5. Conclusions

An insufficient number of repetitions prevents the successful deciphering of the data by the receivers, leading to a high bit error rate. Excessively high repetitions of UE pieces lead to the wastage of valuable wireless resources in 3GPP mMTC. Therefore, in the present study, adaptive repetition approaches with machine learning were developed to substantially increase the network transmission efficacy for eMTC systems in mMTC. The simulation results show that the proposed RLA could effectively improve the transmission probabilities, the resource utilization, the average number of repetitions, and the average energy consumption. The proposed RLA is more suitable than the common LUT for the eMTC system in mMTC. In future work, we will adopt online deep learning approach for 6G or open radio access network (O-RAN) AI architecture and specific use cases. We will enable the approach to learn effectively in special communication scenarios with difficult-to-obtain training samples and have more appropriate depth and accuracy of learning.

Author Contributions: Conceptualization, L.-S.C., C.-H.H., C.-C.C., Y.-S.L. and S.-Y.K.; Writing—original draft, L.-S.C. and C.-H.H.; Writing—review & editing, C.-C.C., Y.-S.L. and S.-Y.K. All authors have read and agreed to the published version of the manuscript.

Funding: The original research work presented in this paper is partly related to BSMI under contract number 1D201101221-134 awarded by BSMI Taiwan and MOST under contract 111-2218-E-305-002-.

Institutional Review Board Statement: Not applicable.

Acknowledgments: The authors would like to thank the BSMI (Bureau of Standards, Metrology and Inspection, Taiwan) research group and MOST for their technical support.

Conflicts of Interest: The authors declare no conflict of interest.

References

1. Service Requirements for Machine-Type Communications (Release 17), Document 3GPP TS 22.368 V17.0.0, April 2022. Available online: https://www.etsi.org/deliver/etsi_ts/122300_122399/122368/17.00.00_60/ts_122368v170000p.pdf (accessed on 18 October 2022).
2. Verma, R.; Prakash, A.; Agrawal, A.; Naik, K.; Tripathi, R.; Khalifa, T.; Alsabaan, M.; Abdelkader, T.; Abogharaf, A. Machine-to-machine (M2M) communications: A survey. *J. Netw. Comput. Appl.* **2016**, *66*, 83–105. [CrossRef]
3. Qipeng, S.; Nuaymi, L.; Lagrange, X. Survey of radio resource management issues and proposals for energy-efficient cellular networks that will cover billions of machines. *EURASIP J. Wirel. Commun. Netw.* **2016**, *2016*, 140.
4. Bockelmann, C.; Pratas, N.K.; Wunder, G.; Saur, S.; Navarro, M.; Gregoratti, D.; Vivier, G.; de Carvalho, E.; Ji, Y.; Stefanovic, C.; et al. Towards massive connectivity support for scalable mMTC communications in 5G networks. *IEEE Access* **2018**, *6*, 28969–28992. [CrossRef]
5. Mahjoubi, A.E.; Mazri, T.; Hmina, N. NB-IoT and eMTC: Engineering results towards 5G/IoT mobile technologies. In Proceedings of the 2018 International Symposium on Advanced Electrical and Communication Technologies (ISAECT, Rabat, Morocco, 21–23 November 2018; pp. 1–7.
6. 5G; Study on Scenarios and Requirements for Next Generation Access Technologies (Release 17), Document 3GPP TR 38.913 V17.0.0, April 2022. Available online: https://standards.iteh.ai/catalog/standards/etsi/17425546-b012-499b-8730-15e038e818 6d/etsi-tr-138-913-v17-0-0-2022-05 (accessed on 18 October 2022).
7. El Mahjoubi, A.; Mazri, T.; Hmina, N. NB-IoT,eMTC, eMIMO and Massive CA: First Africa Engineering Experimental Results Towards the Delivery of 5G/IoT Smart Technologies Applications. *J. Commun.* **2019**, *14*, 216–222. [CrossRef]
8. Service Accessibility (Release 17), Document 3GPP TS 22.011 V17.5.0, December 2021. Available online: https://www.etsi.org/deliver/etsi_ts/122000_122099/122011/17.05.00_60/ts_122011v170500p.pdf (accessed on 18 October 2022).
9. Physical Channels and Modulation (Release 17), Document 3GPP TS 36.211 V17.0.0, January 2022. Available online: https://portal.3gpp.org/desktopmodules/Specifications/SpecificationDetails.aspx?specificationId=2425 (accessed on 18 October 2022).
10. Radio Resource Control (RRC); Protocol specification (Release 17), Document 3GPP TS 36.331 V17.0.0, April 2022. Available online: https://www.tech-invite.com/3m36/tinv-3gpp-36-331.html (accessed on 18 October 2022).
11. Ali, A.; Hamouda, W. On the cell search and initial synchronization for NB-IoT LTE systems. *IEEE Commun. Lett.* **2017**, *21*, 1843–1846. [CrossRef]
12. Kroll, H.; Korb, M.; Weber, B.; Willi, S.; Huang, Q. Maximumlikelihood detection for energy-efficient timing acquisition in NB-IoT. In Proceedings of the 2017 IEEE Wireless Communications and Networking Conference Workshops (WCNCW), San Francisco, CA, USA, 19–22 March 2017; pp. 1–5.
13. Fortino, G.; Gravina, R.; Russo, W.; Savaglio, C. Modeling and Simulating Internet-of-Things Systems: A Hybrid Agent-Oriented Approach. *Comput. Sci. Eng.* **2017**, *19*, 68–76. [CrossRef]
14. Fortino, G. Agents meet the IoT: Toward ecosystems of networked smart objects. *IEEE Syst. Man Cybern. Mag.* **2016**, *2*, 43–47. [CrossRef]
15. Fortino, G.; Russo, W.; Savaglio, C. Simulation of Agent-oriented Internet of Things Systems. In Proceedings of the 17th Workshop" From Objects to Agents, Catania, Italy, 29–30 July 2016; pp. 8–13.
16. Liu, C.H.; Fan, J.; Branch, J.W.; Leung, K.K. Toward QoI and energy-efficiency in Internet-of-Things sensory environments. *IEEE Trans. Emerg. Topics Comput.* **2014**, *2*, 473–487. [CrossRef]

17. Liu, C.H.; Fan, J.; Hui, P.; Wu, J.; Leung, K.K. Toward QoI and energy efficiency in participatory crowdsourcing. *IEEE Trans. Veh. Technol.* **2015**, *64*, 4684–4700. [CrossRef]
18. Ratasuk, R.; Bhatoolaul, D.; Mangalvedhe, N. Performance Analysis of Voice over LTE Using Low-Complexity eMTC Devices. In Proceedings of the 2017 IEEE 85th Vehicular Technology Conference (VTC Spring), Sydney, NSW, Australia, 4–7 June 2017.
19. Böcker, S.; Arendt, C.; Jörke, P.; Wietfeld, C. LPWAN in the Context of 5G: Capability of LoRa WAN to Contribute to mMTC. In Proceedings of the 2019 IEEE 5th World Forum on Internet of Things (WF-IoT), Limerick, Ireland, 15–18 April 2019.
20. Soussi, M.E.; Zand, P.; Pasveer, F.; Dolmans, G. Evaluating the performance of eMTC and NB-IoT for smart city applications. In Proceedings of the 2018 IEEE International Conference on Communications (ICC), Kansas City, MO, USA, 20–24 May 2018; pp. 1–7.
21. Sharma, S.K.; Wang, X. Towards massive machine type communications in ultra-dense cellular IoT networks: Current issues and machine learning-assisted solutions. *IEEE Commun. Surv. Tutor.* **2019**, *22*, 426–471. [CrossRef]
22. Calabrese, F.D.; Wang, L.; Ghadimi, E.; Peters, G.; Hanzo, L.; Soldati, P. Learning Radio Resource Management in RANs: Framework Opportunities and Challenges. *IEEE Commun. Mag.* **2018**, *56*, 138–145. [CrossRef]
23. Abedin, S.F.; Bairagi, A.K.; Munir, M.S.; Tran, N.H.; Hong, C.S. Fog Load Balancing for Massive Machine Type Communications: A Game and Transport Theoretic Approach. *IEEE Access* **2019**, *7*, 4204–4218. [CrossRef]
24. Kafle, V.P.; Fukushima, Y.; Julia, P.M.; Miyazawa, T. Consideration On Automation of 5G Network Slicing with Machine Learning. In Proceedings of the 2018 ITU Kaleidoscope: Machine Learning for a 5G Future (ITU K), Santa Fe, Argentina, 26–28 November 2018.
25. Liu, Z.; Dai, Z.; Yu, P.; Jin, Q.; Du, H.; Chu, Z.; Wu, D. Intelligent station area recognition technology based on NB-IoT and SVM. In Proceedings of the IEEE 28th International Symposium on Industrial Electronics, Vancouver, BC, Canada, 12–14 June 2019.
26. Comsa, I.S.; Domenico, A.D.; Ktenas, D. QoS-driven scheduling in 5G radio access networks—A reinforcement learning approach. In Proceedings of the GLOBECOM 2017–2017 IEEE Global Communications Conference, Singapore, 4–8 December 2017; pp. 1–7.
27. Liu, C.; Wei, Z.; Ng, D.W.K.; Yuan, J.; Liang, Y.-C. Deep transfer learning for signal detection in ambient backscatter communications. *IEEE Trans. Wireless Commun.* **2021**, *20*, 1624–1638. [CrossRef]
28. Li, L.; Ghasemi, A. IoT Enabled Machine Learning for an Algorithmic Spectrum Decision Process. *IEEE IoT J.* **2019**, *6*, 1911–1919. [CrossRef]
29. Reddy, M.P.; Santosh, G.; Kumar, A.; Kuchi, K. Improved Physical Downlink Control Channel for 3GPP Massive Machine Type Communications. In Proceedings of the International Conference on Communication Systems and Networks, Gurgaon, India, 1–2 December 2018; pp. 1–25.
30. Daniels, R.C.; Caramanis, C.; Heath, R.W. A supervised learning approach to adaptation in practical MIMO-OFDM wireless systems. In Proceedings of the IEEE GLOBECOM 2008-2008 IEEE Global Telecommunications Conference, New Orleans, LA, USA, 30 November–4 December 2008; pp. 1–5.
31. Daniels, R.C.; Caramanis, C.; Heath, R.W., Jr. Adaptation in convolutionally coded MIMO-OFDM wireless systems through supervised learning and SNR ordering. *IEEE Trans. Veh. Technol.* **2010**, *59*, 114–126. [CrossRef]
32. Puljiz, Z.; Park, M.; Heath, R.W., Jr. A machine learning approach to link adaptation for SC-FDE system. In Proceedings of the 2011 IEEE Global Telecommunications Conference-GLOBECOM, Houston, TX, USA, 5–9 December 2011; pp. 1–5.
33. Sun, X.-Q.; Chen, Y.-J.; Shao, Y.-H.; Li, C.-N.; Wang, C.-H. Robust nonparallel proximal support vector machine with lp-norm regularization. *IEEE Access* **2018**, *6*, 20334–20347. [CrossRef]
34. Charrada, A.; Samet, A. Nonlinear Complex LS-SVM for Highly Selective OFDM Channel with Impulse Noise. In Proceedings of the International Conference on Sciences of Electronics, Technologies of Information and Telecommunications, Sousse, Tunisia, 21–24 March 2012; pp. 696–700.
35. Zhou, K.; Zhang, L.; Jiang, M. Enhanced effective SNR prediction for LTE downlink. In Proceedings of the IEEE/CIC International Conference on Communications in China (ICCC), Shenzhen, China, 2–4 November 2015.
36. Study on Narrow-Band Internet of Things (NB-IoT)/enhanced Machine Type Communication (eMTC) Support for Non-Terrestrial Networks (NTN), Document 3GPP TR 36.763 V17.0.0, June 2021. Available online: https://www.tech-invite.com/3m36/toc/tinv-3gpp-36-763_a.html (accessed on 18 October 2022).

 *electronics*

Article

GDPR Personal Privacy Security Mechanism for Smart Home System

Yun-Yun Jhuang [1], Yu-Hui Yan [2] and Gwo-Jiun Horng [2,*]

[1] Department of Management Information Systems, National ChengChi University, Taipei 11605, Taiwan
[2] Department of Computer Science and Information Engineering, Southern Taiwan University of Science and Technology, Tainan 71005, Taiwan
* Correspondence: grojium@stust.edu.tw

Abstract: In the era of vigorous development of the Internet of Things (IoT), the IoT has been widely used in people's daily life. Before the user starts using an IoT product, the developer provides a privacy consent form for the user to fill in. However, the content of the consent form is usually too long for the user to read, and the user neglects the provisions related to privacy use, which often results in personal information being recorded in the database of the product without the user's knowledge. To protect users' informed use, we propose a privacy protection standard of the general data protection regulation (GDPR) law applicable to smart-family-related applications and data security with a consensus mechanism. We also propose a unified device data format agreement. Each product can communicate with each other through a smart housekeeper and can collect personal information between its own products and users based on the personal data protection law. Through practice, we demonstrate the feasibility of this open system. In addition, we also collected 70 questionnaires. If the GDPR specification is placed on smart appliances, about 90% of people can accept smart appliances. If smart appliances can be compatible with different brands' unified standards, about 97% of people can accept smart appliances. Therefore, we recommend the introduction of GDPR specifications for smart home appliances.

Keywords: Internet of Things (IoT); EU general data protection regulation (GDPR); consensus mechanism

Citation: Jhuang, Y.-Y.; Yan, Y.-H.; Horng, G.-J. GDPR Personal Privacy Security Mechanism for Smart Home System. *Electronics* **2023**, *12*, 831. https://doi.org/10.3390/electronics12040831

Academic Editor: Christos J. Bouras

Received: 7 January 2023
Revised: 1 February 2023
Accepted: 3 February 2023
Published: 7 February 2023

1. Introduction

With the rapid development of science and technology, the application of the IoT has become an indispensable part of people's lives. According to a McKinsey Digital report, its IoT will have a certain impact on the economy by 2025. From the statistics, it can be seen that the total undervalued value is 4 trillion, and the overvalued value is 11 trillion [1]. Its application can range from small to large scale. For small scale, it can be used to understand and control home appliance status from a distance in terms of general household equipment, such as air conditioners, refrigerators, televisions, etc. The large-scale representative is a large-scale factory with extremely high productivity. With the maturity of the IoT, production speed can be improved to increase economic value.

The convenience of the IoT is beyond human imagination. The most common places to use the IoT are factories and cities. With the emergence of many sensors and Radio Frequency Identification (RFID) technology, factories have become the places where the IoT is applied most, leading to the emergence of the term Industry 4.0 [2]. After the sensor technology is combined with the machine and connected to the Internet, the sensor data are transferred into the database, and the data are analyzed in real time so that the production line can be corrected in the shortest time, thus improving production speed. In addition, it can improve product quality and enable employees to efficiently produce customized products [3].

In order to improve the quality of life, many manufacturers have developed smart home appliances. Generally, smart appliances record the user's habits or judge weather conditions of the day and adjust themselves. For example, the air conditioner will give the most comfortable temperature and wind direction according to the indoor temperature and human body position. However, smart devices are constantly innovating, and they have also developed exclusive applications that can interact with home electronic devices. Some manufacturers have designed a master control system combining artificial intelligence and speech recognition to provide users with voice control to control household appliances.

In real life, the development and practical application of the IoT can connect hundreds of millions of devices [4], indicating the arrival of huge amounts of data, and some of the data are relatively private information, such as username, IP address, time of use, and religious belief. The data generated by IoT devices can be used for a wide range of purposes. The data are likely to be analyzed without the user's knowledge, which can lead to data misuse and victimization of the user.

As a variety of smart home appliances replace traditional home appliances, it simultaneously raises some concerns and has the potential to improve the quality of life of users. Smart home appliances on the market have a variety of contents and services. When users buy products, it means they need to match the applications provided by the manufacturer to interact. Users are constrained by the products designed by the brand, which limits their choices and leads to low use rate of smart home appliances [5]. Users must fill in terms before operating the application. Generally, the consent terms ask users whether they agree to collect personal information. When users choose not to, the application will not operate. Users need to consent to the collection of personal information, which means that personal privacy may be exposed at any time.

Most users in China have not yet understood the personal data collected by devices. To help users understand the details of personal data collection and protect users more simply, this paper joined the most stringent European General Data Protection Regulation (GDPR) in the world, designed the smart home devices in this study to protect users, and paid special attention to the memory block of personal data exposure.

This paper focuses on two key projects, namely, GDPR personal information protection and consensus mechanisms to give users the right to choose their personal information and equipment data generated by smart housekeepers. The goal is to provide users the right to choose their personal information freely. In recent years, GDPR has been the most rigorous data protection law and the focus of attention of all countries, so all countries are following suit. Those who need to enter the EU market must comply with the relevant provisions of the GDPR. In recent years, the cookie consent notice seen by browsing a website is also set according to the GDPR [6]. Because the website may involve the collection of messenger data or tracking the location, users can protect the right to personal sensitive information through this regulation. The consensus mechanism is a technology of memory blockchain, which is one of the most core technologies. It is a mechanism used to ensure that participants reach consensus and achieve trust between blocks through decentralized consensus algorithms. At present, the consensus mechanism is applied in the field of cryptocurrency. Through the consensus mechanism, fairness, efficiency, and consistency can be achieved.

This paper combines the concept of legal protection and consensus mechanisms into IoT technology to implement security management on the data generated by the equipment in the smart butler and the user's personal information data. Through the framework proposed in this paper, users can better understand the personal data processing principles in EU general information protection regulations and their own right of refusal. If users want to remove the device history records stored in the system, they can implement the right of deletion to remove the records.

In this manuscript, we added the GDPR data protection specification to the intelligent butler equipment of the Internet of Things to realize the GDPR data protection specification. In contrast, on the basis of compliance with the principle, GDPR system was not used, the minimum collection volume of GDPR data was kept confidential for the user's personal data, and GDPR pseudonym protection was supported to make it impossible to meet the requirements.

In addition, the user can decide whether the GDPR rejection right of the recorded data needs to be checked according to the individual. Compared with the existing system, users can have more choices in terms of recording personal data.

Compared with the existing service communication architecture and standards in the smart home industry, the main advantage of this research was to propose a unified device data format protocol. Each product can communicate with each other through a smart housekeeper and can keep the personal information collection between its own product and users based on the personal data protection law. Therefore, the protection of personal information is relatively complete.

This paper contributes to the research literature in four major areas: (1) using the unified device data format protocol, each product can converge and transmit information to each other, and each product can maintain data collection with users; (2) we designed and imported GDPR data protection mechanisms into the smart home appliance IoT platform; (3) we increased the lifetime, interaction, and thoroughness of interest groups; and (4) it promoted people's willingness to use the smart family system to realize these goals.

The framework of this paper is mainly divided into five sections. The first section states the background, motivation, and research purpose of this paper, and it outlines the framework of each chapter. The second section discusses the related literature, including IoT, memory blockchain, and GDPR. Through this section, we can better understand the basic concepts of this paper. The third section is the research method of the paper. It presents the overall architecture of the intelligent butler system in the form of a system diagram and then explains how the equipment created in this design can protect the user's personal resources and how to combine IoT equipment with the consensus mechanism in the memory blockchain. The fourth section presents the experimental process and explains in detail where and how to apply GDPR in this system. The fifth section is the summary, contribution, and suggestions for future research.

2. Related Work

This section introduces the relevant content and technology of this paper, which will facilitate the subsequent system introduction, including the Internet of Things, general data protection regulations, and consensus mechanisms.

2.1. IoT

IoT technology serves to connect various independently operated devices or objects to the Internet [7] and realize interconnection and intercommunication. There are two ways for objects to connect to the network: wired networks or wireless networks. The most common way is to connect to wireless networks. Through wireless network technology, not only can the data obtained by devices can be transmitted to computers or servers, but mobile phones or computers can also be used to connect objects to devices or machines for control. In daily life, IoT technology is mostly used in factories, but in recent years, the application of home IoT has gradually become a trend [8].

In the application of home IoT, the most common smart devices include smart light bulbs, smart switches, sweeping robots, and smart speakers. The difference between smart home appliances and general home appliances lies in whether there is an Internet connection. By using the Internet connection method, users can use mobile phones or computers to control home appliances in other places at any time to achieve a system of interconnection between things. The main concept of IoT technology is information reading and transmission. Reading is to obtain information through sensors, while transmission is to transfer information obtained by sensors through the Internet [9].

The concept of IoT originated in 1970. At that time, the world's first IoT device connected to the Internet was a Coke vending machine [10], which was in the Carnegie Mellon University (CMU) in the United States. It was developed by students of the

Department of Computer Science. It provided functions such as confirming the quantity of beverages in the vending machine and checking the inventory.

According to literature records, the term IoT officially appeared in public in 1999, when it was first proposed by Kevin Ashton of Procter&Gamble (P&G). At first, Kevin Ashton used the title [11,12] in his speech to explain how to apply Radio Frequency Identification (RFID) to the company's supply chain. So far, IoT involves many technologies, such as cloud computing, low-energy wireless communication, and wireless sensor networks, and these technologies are also developing continuously [13].

IoT architecture is generally divided into two types, namely, three-tier architecture and five-tier architecture. The most common is three-tier architecture [14]. The three tiers are the perception layer, network layer, and application layer. IoT devices can be transmitted through wireless networks, mobile networks, Bluetooth, or wired networks [15]. Application layer applications cover everything in human life, such as smart homes that can improve the quality of life, smart agriculture that can monitor the quality of crops, and smart cities that can assist medical personnel in medical care and monitoring traffic conditions [16,17].

2.2. GDPR

GDPR [18,19], jointly formulated by the European Parliament, the European Executive Committee, and the European Council, has 99 articles. It was passed in April 2016 and took effect in May 2018 [20–24], replacing the Data Protection Directive launched by the European Union in 1995. GDPR is a regulation on the protection of personal data and privacy of all EU citizens in EU laws. It is implemented in countries belonging to the EU. All enterprises that have business dealings with EU countries, regardless of their location, also belong to the implementation scope of GDPR.

On 14 April 2016, the European Parliament adopted the GDPR, and the regulation came into force 40 days after it was published in the Official Journal of the European Union on 24 May of the same year [25]. On 25 May 2018, two years after the regulation came into force, the EU regulations directly applied to all Member States. On 20 July of the same year, the Joint Commission of the European Economic Area and Iceland, Liechtenstein, and Norway reached an agreement to comply with the regulation, and GDPR came into effect in the countries of the European Economic Area.

The differences between the national individual capital law and GDPR are shown in Table 1. Within the scope of regulation, it is difficult for the country to prosecute and punish overseas offenders due to its international status. The difference between the requirements of consent is that the country can obtain the consent of the data subject explicitly or implicitly, and GDPR must inform the clear action. A vague description or an option that is preset as consent may violate the GDPR. The right to be forgotten in the home country is notified by the processor to the party concerned that the specific purpose of collecting personal information disappears or the party concerned requests to delete personal information.

In addition to the above circumstances, GDPR also gives the party concerned the right to withdraw its consent. The data portability right has no relevant provisions in the country. GDPR stipulates that the data subject has the right to require the data controller to provide itself or transmit it to other designated controllers. Regarding obligations of data controllers and processors, each country requires that the safekeeping of personal data must take security measures and meet the current technological or professional standards. When GDPR requires large-scale processing of personal data, a data protection impact assessment should be made, and a dedicated data protector should be appointed. In principle, the country allows cross-border transmission of individual assets. Except for special circumstances, GDPR prohibits cross-border transmission of individual assets in principle, except for obtaining sufficient recognition or enterprises meeting the protection measures.

Table 1. Differences between China's individual capital method and GDPR [26].

Matter	National Personal Capital Law	GDPR
Scope of specification	It is difficult to prosecute foreigners	If we collect personal resources from EU citizens, it will be regulated
Requirements of consent	Can be expressed or implied	Must inform clear action
The right of data subject to be forgotten	The processor shall take the initiative to inform that the specific purpose of personal data collection disappears, or the party concerned requests to delete personal data	Give the parties the right to withdraw their consent
Data portability of data subject's rights	No relevant regulations	The data subject has the right to require the data controller to provide itself or transmit it to other designated controllers
Obligations of data controllers and processors	Security measures must be taken to keep personal data, and the technology or professional standards at that time must be met	When processing individual resources on a large scale, a data protection impact assessment shall be prepared, and a dedicated data protection officer shall be appointed
Cross-border transmission	Cross-border transmission is allowed and prohibited only under special circumstances	It is only acceptable to obtain sufficient certification or the enterprise complies with the protection measures

2.3. Consensus Mechanism

As one of the core technologies of the memory blockchain [27], the consensus mechanism plays an indispensable role in obtaining protocols in a distributed environment. The consensus mechanism is a combination of consensus and mechanism. The consensus is to agree on different opinions or interests and achieve consistency. The mechanism is a rule. As the memory blockchain is a point-to-point network system, anyone can participate in the network and use the system without a central server to jointly manage the entire system. It is thus necessary to maintain the operation order and fairness of the system by the rules of the consensus mechanism and reward the nodes that provide resources to maintain the memory blockchain and punish the nodes that intend to harm the system.

Most people think that the consensus mechanism is the protocol generated by the memory blockchain, but in fact, the consensus mechanism came out about 20 years earlier than the memory blockchain. The consensus mechanism appeared in 1989. Lynch, Dwork, and Stockmeyer first proposed in 1988 that consensus was the beginning in the case of partial synchronization [28], while the first consensus mechanism was the Paxos algorithm [29] proposed in 1989. Subsequently, the Raft algorithm, Byzantine fault tolerant, and multi-Byzantine protocols were derived. In recent years, with the popularity of memory blockchain cryptocurrencies, many consensus mechanisms suitable for cryptocurrencies have been developed, and each cryptocurrency uses different consensus mechanisms.

At present, there are eight common consensus mechanisms: workload proof [30], holding proof [31], agent holding proof [32], space proof [33], Paxos algorithm, Raft algorithm [34], Byzantine fault tolerance [35], and LibraBFT [36]. None of the eight consensus mechanisms is perfect, and each has its own advantages and disadvantages. Although there is currently no perfect consensus mechanism, there is a concept of cryptocurrency using a hybrid mechanism, which combines workload proof and holding proof to balance their respective shortcomings.

Jingwen Pan et al. compared three main consensus mechanisms in the paper Development in Consensus Protocols: From PoW to PoS to DpoS, which were workload proof, holding proof, and agent holding proof. The author said that newer protocols could solve the problems of previous protocols. For example, proof of holdings and proof of agent holdings could solve the problems of running speed and resource consumption of proof of workload. The security problem also alleviated 51% of attacks harmful to proof of workload [37]. According to Omar Alfandi et al. in [38], in the context of the IoT and memory blockchains, using Byzantine fault-tolerant consensus protocols to select a group of authenticated devices in the network was considered as a more efficient solution than other consensus protocols. In this paper, the authors evaluated the fault tolerance of different network settings and verified their proposed model. The research results showed that the mixed scenario proposed by the authors was better than the non-mixed scenario.

The Byzantine general problem is a distributed peer-to-peer network communication fault-tolerance problem, which was proposed by Leslie Lamport in 1982. In the paper The Byzantine General Problem, the author discussed how a reliable system should handle the failure of one or more computers, and the computer with failure may often be ignored or send wrong conflict information. The author called these problems Byzantine problems [39].

Considering that the personal information collected in smart appliances and the data generated by smart appliances need to be properly and reasonably operated, this paper combines GDPR to regulate the right to use personal information and device data. The user can clearly know the use of personal information and equipment data through the GDPR specification of the system, and the user can also decide whether to accept the data generated by the system's recording equipment. According to the provisions in the GDPR, users need not worry about whether they need to have the protection of this regulation in a specific country.

With the emergence of a large number of smart home appliances, ensuring the safety of equipment is a difficult and very important task. We investigated how to confirm whether the equipment is under the control of the intentional person. Therefore, among many consensus mechanism technologies, the choice of Byzantine general is the most appropriate. We used the concept of Byzantine general problem to judge whether the equipment is safe based on the consensus reached among the equipment terminals.

3. System Model

Due to the maturity of IoT technology and the increasing number of users, personal digital information and values in the living environment have become indispensable data for IoT technology. The more data are obtained by the IoT, the more convenient life will be. In view of data security, this system was standardized with the most stringent GDPR, and the consensus mechanism technology was used to confirm whether the equipment was controlled by unknown people. This paper hopes to solve users' data security concerns through three different technologies.

The system architecture of this paper is shown in Figure 1, which is composed of three parts: GDPR provisions, equipment data format conversion, and the consensus mechanism. GDPR provisions should be applied to the template system to provide a guarantee for users' personal information. When users add smart home appliances, they can decide whether to record device data according to their preferences. Since there is no uniform data format in the smart home system on the market today, if users want to buy smart home appliances, they must choose the same brand, which indirectly leads to a decline in users' purchasing desire. To solve this problem, the system will convert the data format of household appliances of different brands so that users can choose more smart appliances without being limited to the same brand. The consensus mechanism is used to ensure the security of the user's equipment. The Byzantine general problem using the consensus mechanism can determine whether the equipment is under the control of the user.

In this paper, the system operation module is shown as the schematic diagram of each module in Figure 2, which is divided into four modules: server side, device control side, user control side, and consensus mechanism. The server side is responsible for providing services, the device control side is responsible for converting the device data format and communicating with the server side, and the user control side is responsible for providing the interface for the user to control the device. Then this paper introduces the hardware equipment used by the four modules, the server erected, the transmission mode, the technology used, the operation mode provided for users, and how to combine the four modules.

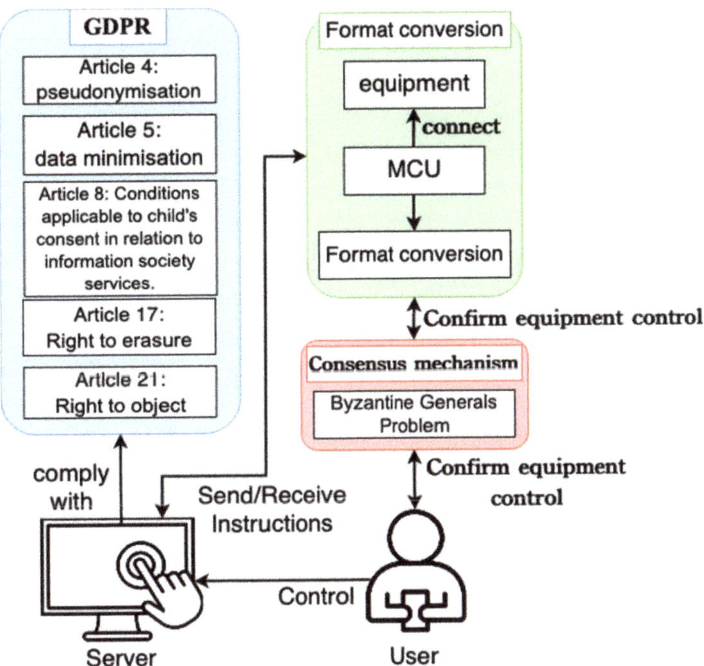

Figure 1. Schematic diagram of system architecture.

The hardware device used on the server side was Raspberry Pi 4, and the socket server, PostgreSQL database, PHP website, and user interface provided for users to operate were set up in Raspberry Pi 4. The socket server also serves an important role as a bridge between the transmission of each module. The device control end and the user control end are connected to the server through the wireless network. The user control end can transmit the instructions to the device control end through the socket and control the household appliances.

The database built on the server side was PostgreSQL, which is an associated database. The database is responsible for storing various data in this system. The data stored include the user account password and the use status of household appliances. The purpose of recording the user account password is to provide users with access to the system, while the purpose of recording the status of household appliances is to provide users with a historical record of the real-time status or status of household appliances.

The server-side user interface is written using Python language graphical interface PyQt, which provides users with a simple operation screen. This service combines GDPR and provides users with six operation functions, including adding users, adding devices, logging off devices, deleting records, viewing device status, and viewing historical records. This paper introduces how to integrate GDPR into each function and the applied provisions in Section 3.2 and describes the provisions in detail.

The main hardware of the device control terminal is the development version of ESP32 single chip microcontroller, which combines Wi-Fi and Bluetooth functions and has a low cost. ESP32 is used to connect with household appliances and convert the data format of the original household appliances to the unified format of the system. ESP32 transmits the converted data to the socket server on the server side through Wi-Fi and stores the home appliance status in the database. The device control end generates a log file and transmits it to other home appliances and the user control end, and it confirms with the user control end through the consensus mechanism to ensure that the device is controlled by family members. In Section 3.3, this paper introduces how the system uses the technology of the consensus mechanism.

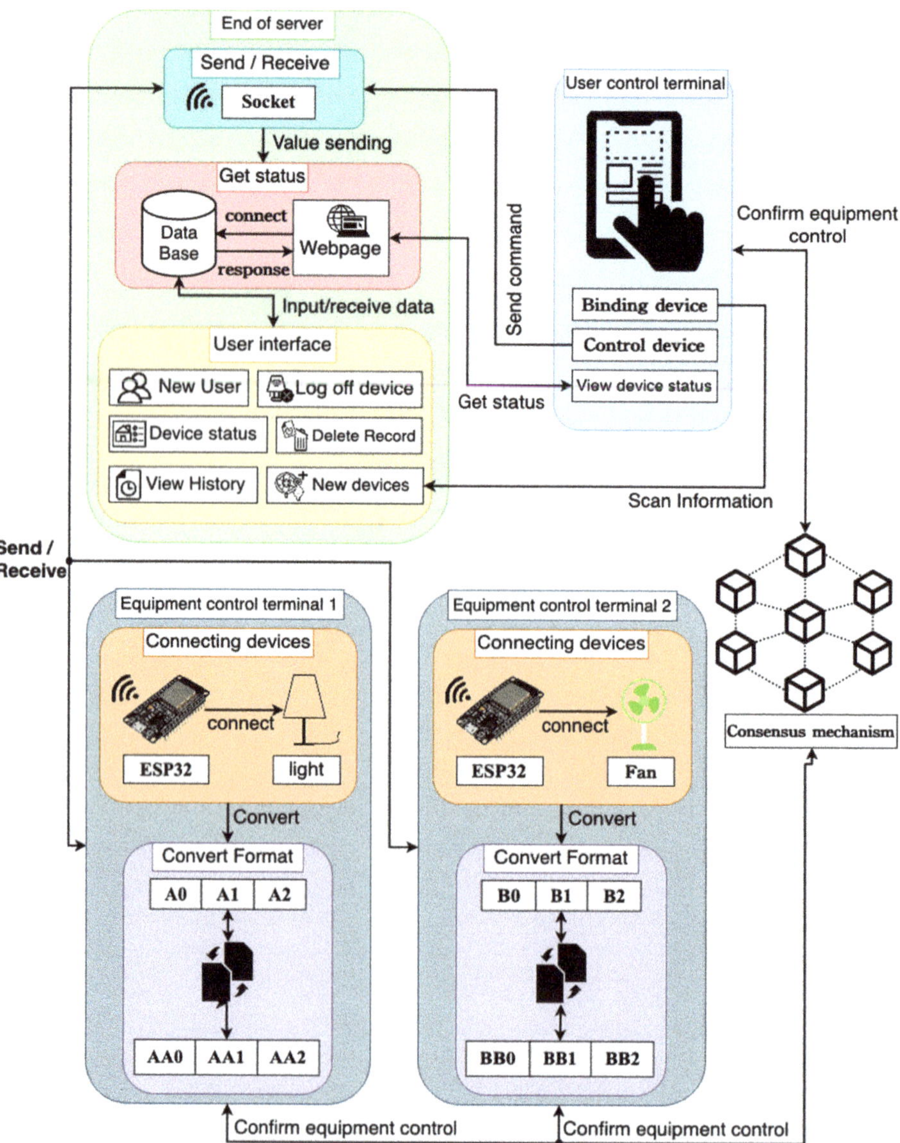

Figure 2. Module connection framework.

The mobile application developed by the user control terminal for this system is written using Flutter suite. The user control terminal is designed with five functions for users to operate, including user login, binding device, controlling device, viewing device status, and logging off. The user login function can only log in to the account that has been applied for on the server to ensure that the user is a family member. The device-binding function scans the QR code generated by the server to bind the device to the APP, and the control device function controls the device. The view device function allows users to view the current usage status of the device.

3.1. GDPR Articles

GDPR is the largest and most rigorous data protection regulation. Compared with other data protection regulations, the scope of GDPR is not only limited to specific regions but also to the people or companies of any country. The systems generated by European companies and the software and hardware that European people will use are subject to the scope of GDPR. This paper used five GDPR provisions to provide users with a secure system environment.

The GDPR provisions used in this system are shown in Table 2. Item 5 pseudonymization in the definition of Article 4 refers to the mechanism of processing personal data. Without the use of additional information, personal information cannot be identified, provided that the additional information is kept separately and subject to the constraints of science, technology, and organizations to ensure that the personal information cannot identify the person concerned. The system complies with this provision and changes the identifiable name to the name filled in by the user's preference so that the information of the party concerned cannot be identified.

Table 2. Explanation of GDPR Articles [40].

Article No	Article Name	Article Description
Article 4 Item 5	Pseudonymization	It means that personal information cannot be identified without using additional information.
Article 5 Point c of Item 1	Principle of minimum data collection	Personal data should be appropriate, relevant, and limited to the minimization of necessary data related to processing purposes.
Article 8	Conditions applicable to child's consent in relation to information society services	It is legal to process the personal information of children over the age of 16, but it is legal only with the consent and authorization of the legal guardian if they are under the age of 16.
Article 17	Right to be forgotten	The data party shall have the right to delete personal information from the controller without unreasonable delay, and the controller shall have the obligation to delete personal information.
Article 21	Right to object	When the controller processes personal data for direct marketing purposes, the data party has the right to refuse to process the personal data involved in marketing purposes at any time.

In addition to the inability to identify the information of the person concerned, the principle of minimum data collection in point c of Item 1 of Article 5 (Personal Data Processing Principles of the Regulations) shall also be observed, which means that the personal data shall be appropriate, relevant, and limited to the minimization of necessary data related to processing purposes.

Article 8 of the regulation refers to the conditions for children's consent in information society services, which are divided into three items. Item 1: if a child is 16 years old or older, it is legal to process the child's personal information, but if the child is not 16 years old, such processing must be authorized by the consent of the legal guardian. However, EU Member States can define a lower age for passing the law, provided that the minimum age is not less than 13 years old. Item 2: under the available technology, the controller shall make reasonable efforts to verify whether the legal representative agrees or authorizes. Item 3: (1) shall not affect the general regulations of EU Member States, such as the provisions on the validity, formation, or impact of regulations related to children.

Article 17, right to erasure, is also called right to be forgotten, and its provisions are divided into three items. Item 1: when personal information is no longer needed or is illegally processed, data parties shall have the right to obtain the deletion of personal information from the controller without unreasonable delay, and the controller has the obligation to delete personal information. Item 2: if the controller discloses personal information, according to Item 1 of this article, and if the data party requires deletion, the controller shall delete any connection or copy related to the data. Items 1 and 2 of these articles do not apply when Item 3 is to exercise the right to freedom of expression and information or to establish, exercise, or defend legal requirements.

The right to object in Article 21 of GDPR is divided into six items. The first item expressly stipulates that the data party has the right to refuse the regulations based on specific circumstances. Point e or f of the first item deals with relevant personal information, including all filing of this provision, unless the controller proves that the processing was prior to the legal basis, establishment, exercise, or defense of the rights and freedoms of the data party. Otherwise, the controller shall not process personal data. Item 2: when personal information is processed for direct marketing purposes, the data parties have the right to reject the scope of data processing involved in marketing purposes at any time.

Item 3 of the right of refusal is as follows: when the data party refuses to process for direct marketing purposes, such purpose processing will no longer occur. Item 4: when communicating with the data parties for the first time, the rights of Items 1 and 2 shall be clearly put forward, and the difference between any information shall be clearly introduced. Paragraph five states that in the process of using information society services, despite the provisions of Directive 2002/58/EC, data parties may refuse to use the automated methods of technical specifications. Paragraph six states that if the processing of personal data is for scientific, historical, or statistical research purposes in accordance with paragraph one of Article 89, the data party shall have the right to refuse the processing of relevant personal information.

In this paper, the server side of Figure 2 is detailed in Figure 3. The system combines the five GDPR clauses in Table 2 with the server-side user interface functions, as shown in Figure 3. On the server side, there are three kinds of running functions: sending and receiving instructions, fetching status, and user interface. The socket server is used for sending and receiving instructions. The user control end transmits the sent instructions to the socket server. At this time, the socket server transmits the received instructions to the device control end and transmits the device control status to the database for recording. The crawl status function displays the information in the database through the web page.

The user interface function provides six functions, including adding users, adding devices, logging off devices, deleting records, viewing device status, and viewing history. When adding users and new devices, the system writes the new results to the database. Logging off the device and deleting the record removes the corresponding data in the database. Viewing the device status and viewing the history displays the data in the database. Among them, new users, new devices, cancelled devices, and deleted records are combined with GDPR provisions into the function.

When new users are added, the family may include minors, so the age judgment function was added to protect minor children. This design needs to ask whether users are 13 years old or older, and if they are at least 13 years old, it must ask whether they are 16 years old or older. This step is to comply with the conditions for the consent of children involved in information society services. After completion, new users can be added.

The newly added equipment functions should comply with the principle of pseudonymization, minimum data collection, and the right of refusal. Kana is used to record the name of the device, which is customized by the user. In terms of data collection, this paper refers to [41] to summarize the personal information most frequently collected by suppliers, and the data collected by this system are described in Table 3. As shown in Table 3, the system follows the principle of minimum data collection, as it collects equipment status information but excludes equipment values. The right of refusal is provided to the user to refuse to collect data, and the user can decide whether to accept the data collected by the system according to his own investigation.

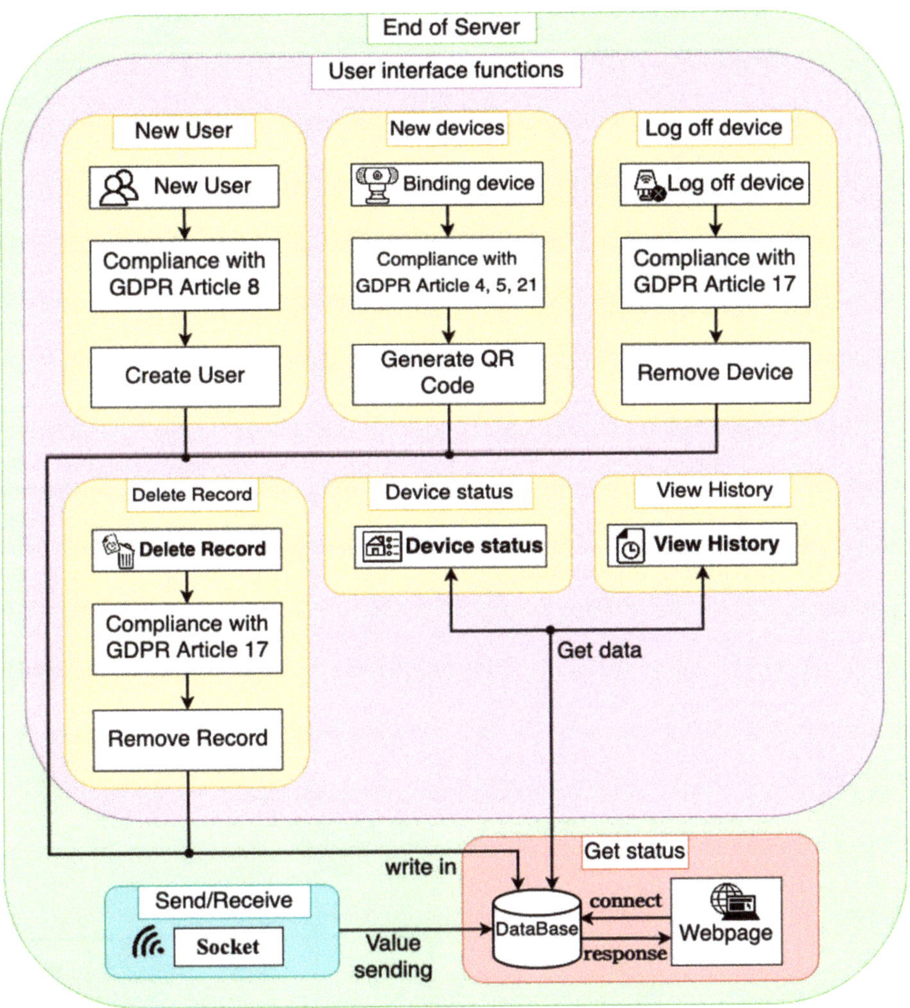

Figure 3. Schematic diagram of server module.

Table 3. Data collection.

	Item	Collect	Do Not Collect
1.	Equipment use status	✓	
2.	Equipment service time	✓	
3.	User email	✓	
4.	Equipment serial number	✓	
5.	Equipment name	✓	
6.	Position		✓
7.	Personal preference		✓
8.	User name		✓
9.	ID card No		✓
10.	Card number		✓

The system has the right to delete the contents of the clauses. When the user needs to remove the equipment or wants to remove the collected information, he/she can perform two functions at any time, namely, logging off the device and deleting the record. If the user executes this function, he/she cannot view the historical record or query the device information. In other words, if deleted, it will be deleted together with the data accessed at that time in the database, leaving no records.

3.2. Equipment Data Format Conversion

Data format refers to the format of data storage records or files of hardware devices. Generally, the types of formats are numerical, binary, octal, or hexadecimal. However, the data formats used by various hardware equipment manufacturers on the market are different. When users buy products from different manufacturers, they need to use the application programs developed by the manufacturers themselves, which causes inconvenience to users.

In order to provide users with multiple choices for household appliances, the system performs data format conversion for household appliances. Figure 4 shows a schematic diagram of the conversion of binary data to the hexadecimal format. Since the device formats provided by various manufacturers are different, it is necessary to connect the home appliance with ESP32 first. After ESP32 connects with the home appliance, it reads the data format of the home appliance and uniformly converts the data to the hexadecimal format through the program.

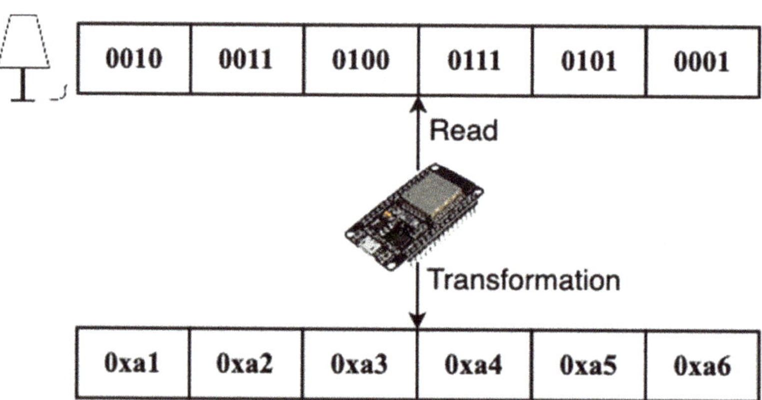

Figure 4. Schematic diagram of converting binary data to the hexadecimal format.

After the device data format is converted by ESP32, the wireless network can be used to transmit packets to the server through the socket. The server confirms the identity of the device by the header and footer of the received packets. As shown in the packet information diagram in Figure 5, the system distinguishes packets of household appliances of different brands by different headers and tails. If the header of packets received by the server is 0xa1 and the tail of packets is 0xa5 and 0xa6, the household appliances can be determined to be A-brand LED.

Except for the header and footer, the rest of the packet is the device information. When the status of the appliance changes, the device control terminal writes the information into the packet, and the server knows the appliance status through the middle section of the packet. The user control end uses the same principle to write the instruction to be controlled into the packet and transmits the packet to the device control end through the socket server. The device control end will change the status according to the received packet.

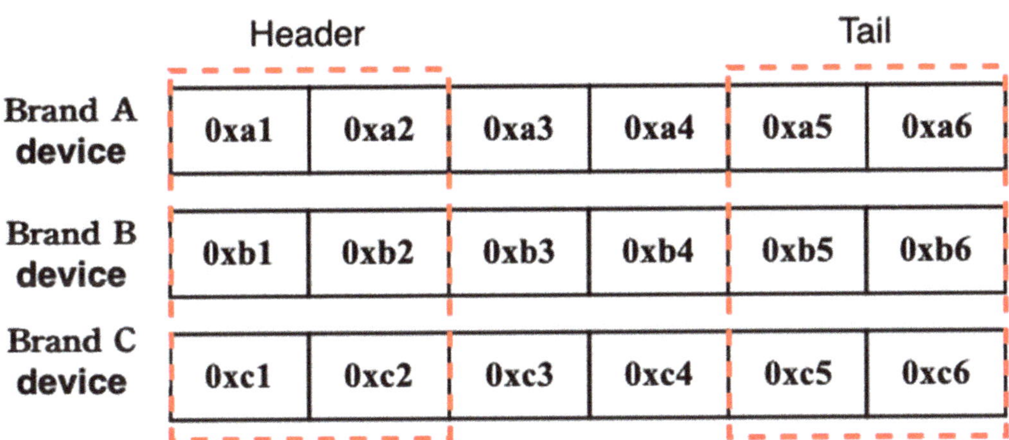

Figure 5. Schematic diagram of packet information.

3.3. Consensus Mechanism

In modern life, the IoT is widely used and convenient. However, with the increasing number of IoT devices, determining how to ensure data security is a major challenge for developers. The security risk of the IoT not only relates to data theft but also to the use various methods to maliciously attack or control appliances. If developers ignore IoT security issues, it may bring inconvenience or danger to users' lives.

For IoT security issues, the system adds consensus mechanism technology to improve security and to quickly ensure that devices are controlled by people with a mind. At present, there are many kinds of consensus mechanisms, each of which has its own advantages, disadvantages, and different functions. The system uses the concept of the Byzantine general problem to reach a consensus between each device control end and the user control end. If one node does not match the information of other nodes, it can be determined that the device is controlled by someone with a mind.

The schematic diagram of the consensus mechanism module of the system is shown in Figure 6. When the user control terminal operates device control terminal two, the user control terminal transmits a log file to device control terminal one, device control terminal two, and device control terminal three, and the operated device control terminal two also transmits the log file to device control terminal one, device control terminal three, and the user control terminal for confirmation. The contents of the log file are the MAC address, the changed state of the equipment, and the time of the changed state of the equipment. If a hacker invaded device control terminal one and operates it, device control terminal one would transmit the log file to device control terminal two, device control terminal three, and the user control terminal, while other terminals would not send out the log file. At this time, it can be known that the operation of device control terminal one is not performed by a family member.

Algorithm 1 shows that the user control terminal operates the virtual code of device control terminal two. The user control terminal writes the MAC address of the device to be controlled, the status to be changed, and the execution time to the log file and sends the log file to all device control terminals. The device control end being operated writes the MAC address, changed status, and execution time to the log file and sends them to the other device control ends and user control ends.

Algorithm 1 Pseudocode of user control terminal operating device control terminal 4

1.	let user = User control terminal
2.	let device_1 = Equipment control terminal 1
3.	let device_2 = Equipment control terminal 2
4.	let device_3 = Equipment control terminal 3
5.	let user_log = Log file of user control terminal
6.	let device_2_log = Log file of equipment control terminal 2
7.	
8.	# User-operated equipment 2
9.	User sends command to device 2
10.	user_log = MAC address, status to be changed, and execution time of the device 2
11.	Send user_ Log to device 1, device 2, and device 3
12.	
13.	if device_2 Receive instructions
14.	device_2 Implementation status change action
15.	Device_2_log = MAC address, change status, and execution time of device 2
16.	Send device_2_Log to device_1, device_3, and user
17.	else
18.	Do not perform actions

The device control terminal judges whether the operation is a family member virtual code, as shown in Algorithm 2. When device control terminal one receives the log files of the user control terminal and device control terminal two, it first judges whether the information of the log files of the user control terminal and the device control terminal are consistent. If the information of the two files is consistent, it can be determined that the operation is conducted by a family member. If the information is inconsistent, it can be determined that the user control terminal is not operated by a family member. If device control terminal one only receives the log file of device control terminal two, it can be determined that the operation of device control terminal two was performed by a hacker.

Algorithm 2 Virtual code for judging operation result at equipment control terminal

1.	let user = User control terminal
2.	let device_1 = Equipment control terminal 1
3.	let device_2 = Equipment control terminal 2
4.	let user_log = Log file of user control terminal
5.	let device_2_log = Log file of equipment control terminal 2
6.	
7.	if device_1 Received user_ Log and device_ 2_ log
8.	# Check the log content
9.	If user_log == device_2_log
10.	Decide to act as a family member
11.	else
12.	User is not a family member
13.	else if Only device is received_ 2_ log
14.	Confirm that the device has been hacked
15.	else
16.	Waiting to receive

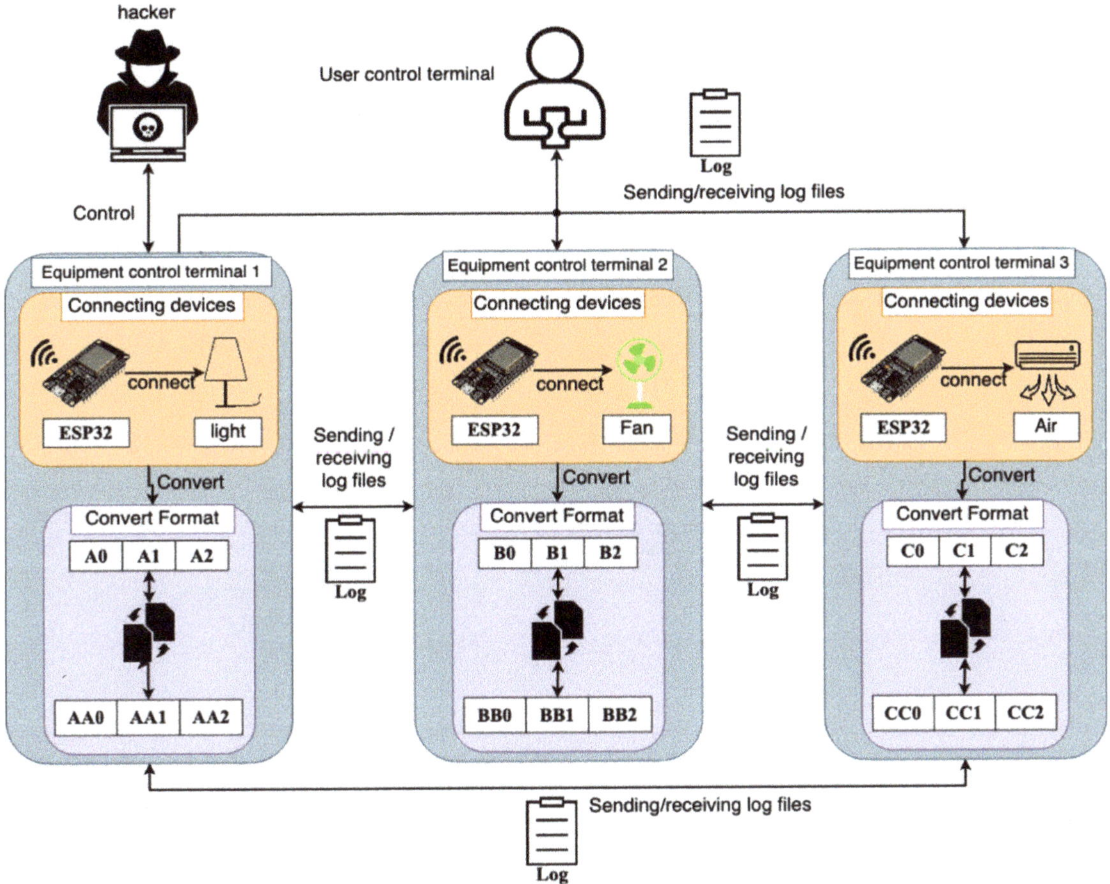

Figure 6. Schematic diagram of consensus mechanism module.

4. Research and Analysis

This paper combined GDPR with IoT technology applications and provided a user interface and user interaction in the server. Figure 7 shows the operation sequence between the user and the server. First, the user creates a new user in the server. During the process of creating the user, the server asks the user about the GDPR clause, and the user replies to the clause. After the server confirms that the user's account format is correct and the fields are filled in, the new user is successfully created.

After the user logs in with the newly created user account, the server checks whether the account exists in the database and then confirms whether the password is correct. If both are correct, the system will jump to the interface for entering the verification code. Then the server randomly generates a group of verification codes. The server writes the verification codes into an email and sends a letter to the user's mailbox. The user needs to enter the verification code in the letter into the field. After the server compares the correctness of the verification code entered by the user, the interface jumps to the control interface.

The user operation's initial screen is shown in Figure 8. This is the user operation interface, which provides three functions: user login, account application, and password forgetting. To improve security, the system limits the user account format to the email format, uses email to verify whether the user is himself or herself, and prevents the account from being stolen by intentional persons. The system will hide the password in the password input field, providing an environment for users to enter the password safely.

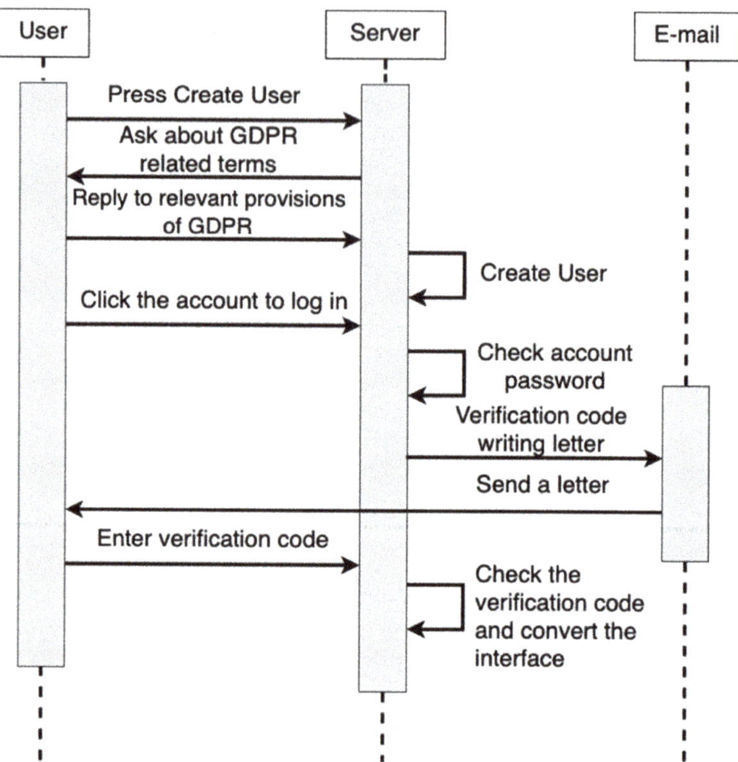

Figure 7. Sequence diagram of user and server operation.

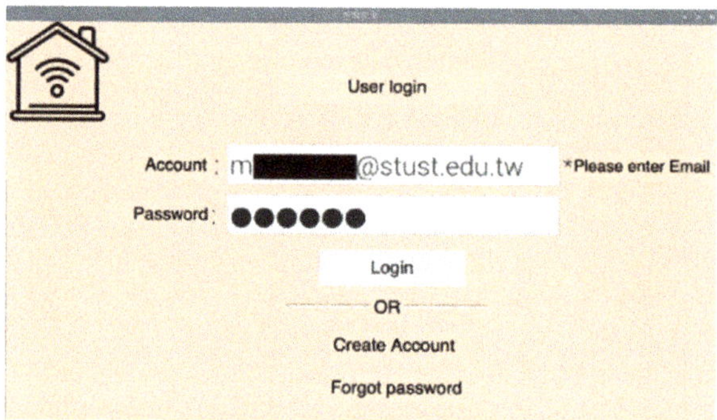

Figure 8. User operation interface.

The account application function abides by Article 8 of GDPR: comply with the consent conditions that involve information society services for children. In Article 8, it is necessary to first pay attention to whether the user's age is 13 years old, as shown in Figures 9 and 10, because it is stipulated that the minimum age is 13 years old. As shown in Figure 9, when the user clicks the option under 13 years old, the system will send it directly without asking for more details and will unconditionally not collect any relevant information from the user in accordance with the provisions. As shown in Figure 10, if the user clicks the option of 13 years old or older, the system will pop up the option of asking whether the user is 16 years old or older.

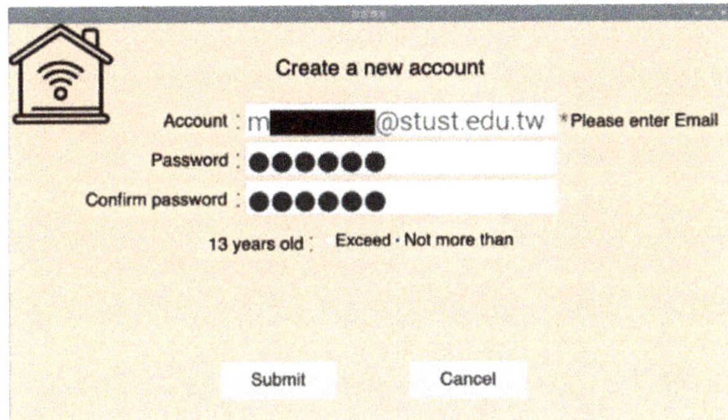

Figure 9. Interface of account creation under 13 years old.

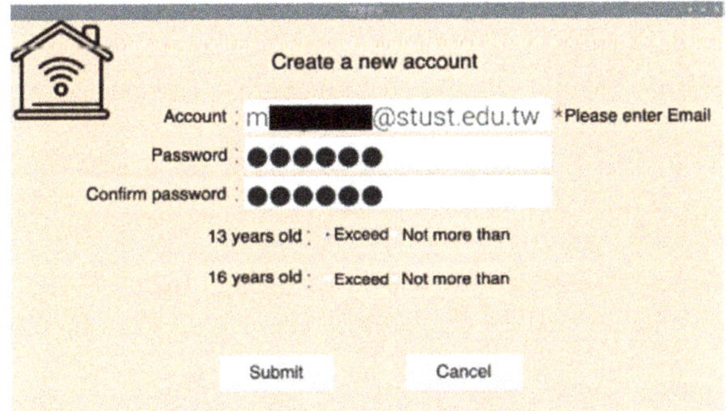

Figure 10. Interface of account creation age over 13.

According to Item 1 of Article 8 of the GDPR, it is legal to process the personal information of users who have reached the age of 16. If the service provider needs to collect the personal information of users who have not reached the age of 16, it is legal only when authorized or agreed to by the legal representative. As shown in Figure 11, when the user indicates that he/she is under the age of 16, the system will pop up and fill in the legal representative email field to be authorized by the legal representative. To comply with the provisions of Item 1, Article 8 of the GDPR, the user must fill in the legal representative email before creating. As shown in Figure 12, if the user clicks the option of over 16 years old, the user can decide the use of personal information by himself/herself without the authorization of the legal representative.

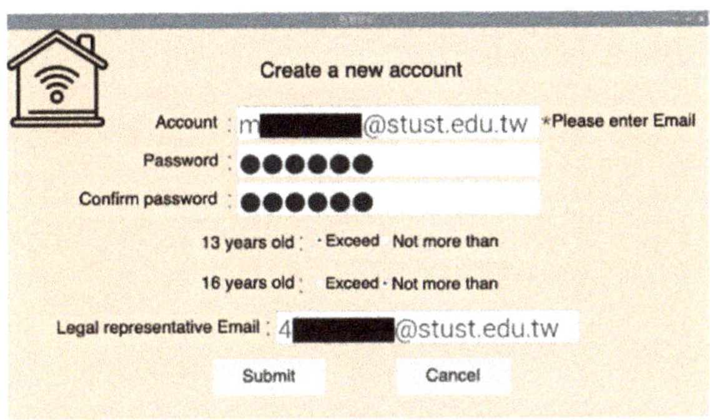

Figure 11. Interface of account creation under the age of 16.

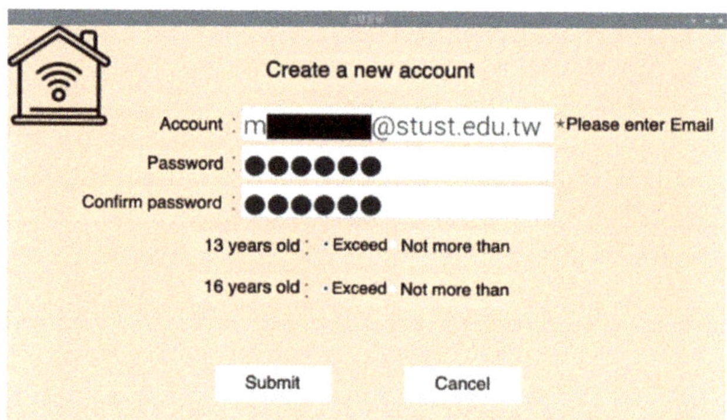

Figure 12. The interface of account establishment when the age is over 16.

After being inquired by the system, the user can create a new user account and log into the user operation interface from this account. The user needs to input the applied account password into the field, and the system will confirm whether the account exists and whether the password is correct in sequence. After confirmation, the system jumps to the screen of entering the verification code and generate a group of verification codes. This verification code is sent to the user's letter by email. The user needs to fill the verification code in the letter into the field. After the system confirms that the verification code is correct, it can jump to the user control interface.

If the user enters the verification code input interface but has not received the letter for a long time, the user can click the re-send verification code function, and the system will re-generate a set of verification codes and send them to the user's email. This system uses the verification code mechanism to send by email. To ensure that the user is himself, even if someone embezzles the account, the user can find out from the verification code letter and change the password.

Then this paper introduces the user operation interface, which provides five main functions and convenient viewing time for users. Four GDPR clauses are combined in the functions of adding devices, canceling devices, and deleting records. Users can decide the access and use of personal information under the protection of the clauses.

The new device functions need to be matched with the webcam. In order to unify the format of household appliances, the system needs to obtain the information of the original household appliances first. First, the user can scan the QR code of the original home appliance. Then the system will capture the required part of the scanned information, such as the manufacturer's license, device name, and device serial number, and then display the captured QR code information in the device information interface of the new device.

This system combines Article 4 (pseudonymization), Article 5 (principle of minimum data collection), and Article 21 (right of refusal) of GDPR. In combination with Article 4, the system provides users with the ability to name household appliances according to their own preferences so as to prevent data information from being unrecognizable when it is stolen. As the system only obtains the manufacturer's brand, equipment name, and equipment serial number (but not other information), it complies with Article 5. The user is asked whether to record the historical status. If the user does not want to record, the provisions of Article 21 can be implemented. After the user fills in the information, the system will generate a new QR code, which will be used to bind the device at the user control end.

The deletion right in Article 17 of GDPR is combined in the function of canceling the device and deleting the device. When the user can delete the device or the use record of the device he wants to cancel, he can implement Article 17 to cancel or delete the device at any time. When indicating the device to be logged off or the record to be deleted, the system will pop up the option of reconfirmation. In order to prevent the user from accidentally clicking, the user needs to enter a password to ensure that the user can only log off or delete the record if it was not accidentally clicked.

This function of viewing device status interface is used to provide users with information about the device. In this paper, LED (LED small bulb) and FAN (motor fan) were used as experimental equipment. The nickname refers to the name that the user chooses for the device. The serial number is the original serial number of the appliance. Whether or not to record the status selected by the user is whether or not to record the status. If the user selects no, the device status cannot be known in this interface. The status is the current use status of the device.

Users can query the status and time of equipment changes through the view history function. In this paper, LED (LED small bulb) and FAN (motor fan) were used as experimental equipment. LED has two states, namely, on and off. The FAN has four states, namely, close, small, middle, and big. In addition to the user operation interface on the server side, the system also provides the user control side and the web page version to view the history. Users can use computers or mobile phones to query the history of the device at home without going to the device to operate.

The user control end of this system is a self-developed mobile application using Flutter as the framework and Dart as the program language. Figure 13 shows the user control terminal login sequence. When the user enters the account and password at the device control terminal to log in, the device control terminal confirms to the server whether there is such an account. After confirming that the account exists, the server confirms whether there was an error in entering the password. After confirmation, the server returns the result to the user control terminal.

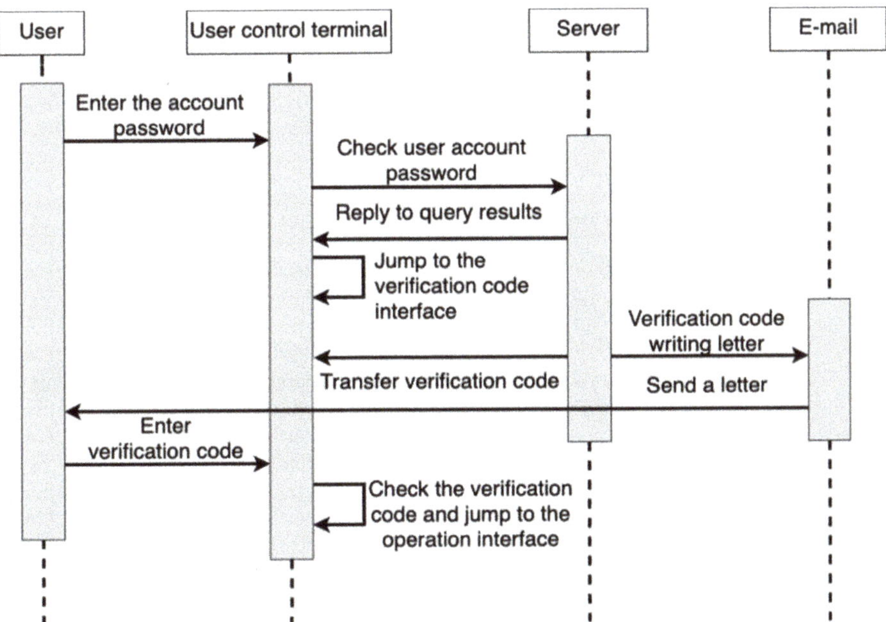

Figure 13. Sequence diagram of user control terminal login.

Currently, the user control terminal performs interface conversion, and the server generates a group of verification codes and sends them back to the user control terminal. At the same time, the verification codes are sent to the user email. The user needs to input the verification code in the email into the verification code interface. After the verification code is identified as correct through comparison, the user control terminal will convert the interface.

A sequential diagram of the user's operation on the device control terminal is shown in Figure 14. After the user presses the device start button (device control terminal one), the user control terminal first sends the command to the server, and the server sends the device start command to device control terminal one after receiving the command. After receiving the command, device control terminal one starts the device and transmits the log file of device control terminal one to the user control terminal and device control terminal two. The log file of the user control terminal is also transmitted to device control terminal one and device control terminal two. Currently, device control terminal two needs to confirm the log files of the user control terminal and device control terminal one.

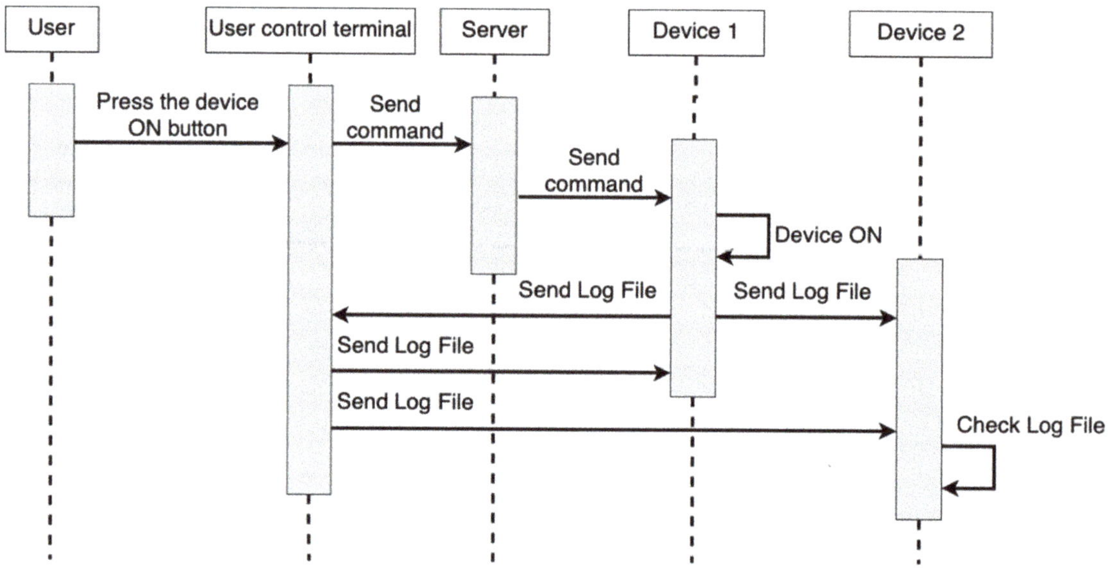

Figure 14. Sequence diagram of user control equipment.

In addition, we collected questionnaires from 70 people, including 34 people aged 16–30 and 36 people aged 31–65. A total of 67 people wanted to buy smart home appliances, 52 people wanted to buy smart home appliances of different brands, and 59 people believed that if different brands are not compatible, it would indeed affect their willingness to use smart home appliances. However, nearly 32 people could not accept smart home appliances to record personal information. However, if the GDPR specification was put on smart home appliances, about 90% of people could accept smart home appliances. If smart home appliances could be compatible with a unified format with different brands, about 97% of people could accept smart home appliances. Therefore, we recommend smart system products. If they are popular with the public, we suggest that the GDPR specification and the transmission formats of different brands can be unified and introduced into smart system products, which will make more people willing to accept them.

5. Conclusions

This paper adds GDPR data protection specification to IoT intelligent housekeeper equipment in order to achieve GDPR data protection specification. Compared with cases where the GDPR system is not used, this system, based on complying with the principle of minimum collection of GDPR data, keeps the user's personal data confidential and protected by means of pseudonymization of GDPR, making it impossible for interested persons to identify the data subject when stealing data. In addition, the user can decide whether the data needs to be recorded according to personal inspection through the GDPR refusal right. Compared with the existing system, the user has more choices in terms of recording personal data.

Compared with the existing service communication architecture and standards in the smart home industry, the main advantage of this research was to propose a unified device data format protocol. Each product can communicate with each other through a smart housekeeper and can keep the personal information collection between its own product and users based on the personal data protection law. Therefore, the protection of personal information is relatively complete. In addition, we also proposed a consensus mechanism to ensure the security of the user's equipment. Through the Byzantine general problem method, we can determine whether the equipment is controlled by the owner. In this study, the concept of consensus mechanisms was used as the protection judgment standard for equipment safety. Through the consensus mechanism, each device end generates an independent log file. When a malicious person intrudes, the user receives the log data of the intrusion device, but it does not send the relevant operation log information. In this way, users can know the important information about the intrusion of the device so that they can take corresponding measures at the first time.

This paper contributes to the research literature in four major areas: (1) using the unified device data format protocol, each product can converge and transmit information to each other, and each product can maintain data collection with users; (2) designed and imported GDPR data protection mechanisms into the smart home appliance IoT platform; (3) increased the lifetime, interaction, and thoroughness of interest groups; and (4) promoted people's willingness to use the smart family system to realize these goals.

Author Contributions: Conceptualization, G.-J.H.; methodology, Y.-H.Y. and Y.-Y.J.; software, Y.-H.Y.; validation, Y.-H.Y., G.-J.H. and Y.-Y.J.; investigation, Y.-H.Y.; resources, G.-J.H.; writing—original draft preparation, Y.-H.Y. and Y.-Y.J.; writing—review and editing, G.-J.H. and Y.-Y.J.; supervision, G.-J.H.; project administration, G.-J.H. All authors have read and agreed to the published version of the manuscript.

Funding: This research received no external funding.

Data Availability Statement: Not applicable.

Acknowledgments: This work was supported in part by the National Science and TechnologyCouncil (NSTC) of Taiwan under Grant NSTC 111-2622-E-218-005- and in part by the Allied Advanced Intelligent Biomedical Research Center, STUST from Higher Education Sprout Project.

Conflicts of Interest: The authors declare no conflict of interest. The authors declare that they have no known competing financial interests or personal relationships that could have appeared to influence the work reported in this paper.

References

1. Manyika, J.; Chui, M.; Bisson, P.; Woetzel, J.; Dobbs, R.; Bughin, J.; Aharon, D. *Unlocking the Potential of the Internet of Things*; McKinsey Global Institute: New York, NY, USA, 2015.
2. Stăncioiu, A. The fourth industrial revolutionindustry 4.0. *Fiabil. Și Durabilitate* **2017**, *1*, 74–78.
3. SEMI Taiwan. *Industry 4.0, Understand It from Shallow to Deep!* Available online: https://www.semi.org/zh/blogs/technology-trends/industry-4.0 (accessed on 31 May 2022).
4. Biswas, A.R.; Giaffreda, R. IoT and cloud convergence: Opportunities and challenges. In Proceedings of the 2014 IEEE World Forum on Internet of Things (WF-IoT), Seoul, Republic of Korea, 6–8 March 2014; pp. 375–376. [CrossRef]
5. Taiwan Network Information Center. 2019 Taiwan Network Report. Available online: https://www.twnic.tw/doc/twrp/201912e.pdf (accessed on 11 July 2022).
6. Top Service Group. Do you Agree to Use Cookies for Tracking? Comply with the EU GDPR Cookie Policy. Available online: https://www.tsg.com.tw/blog-detail3-200-0-gdpr-2.htm (accessed on 10 July 2022).
7. Weber, R.H. Internet of things—Need for a new legal environment? *Comput. Law Secur. Rev.* **2009**, *25*, 522–527. [CrossRef]
8. Mainetti, L.; Mighali, V.; Patrono, L. A location-aware architecture for heterogeneous building automation systems. In Proceedings of the 2015 IFIP/IEEE International Symposium on Integrated Network Management (IM), Ottawa, ON, Canada, 11–15 May 2015; pp. 1065–1070. [CrossRef]
9. OOSGA. IoT Internet of Things—Definition, Application Fields and Actual Industrial Cases. Available online: https://zh.oosga.com/iot/ (accessed on 15 December 2020).

10. The "Only" Coke Machine on the Internet. Available online: https://www.cs.cmu.edu/~{}coke/history_long.txt (accessed on 5 June 2022).

11. Shrouds of Time: The History of RFID. Available online: https://www.railwayresource.com/company/732913/whitepapers/2291/shrouds-of-time-the-history-of-rfid (accessed on 5 June 2022).

12. Ashton, K. That 'Internet of Things' Thing. Available online: https://www.rfidjournal.com/that-internet-of-things-thing (accessed on 5 June 2022).

13. Li, S.; Da Xu, L.; Zhao, S. The internet of things: A survey. *Inf. Syst. Front.* **2014**, *17*, 243–259. [CrossRef]

14. Al-Qaseemi, S.A.; Almulhim, H.A.; Almulhim, M.F.; Chaudhry, S.R. IoT architecture challenges and issues: Lack of standardization. In Proceedings of the 2016 Future Technologies Conference (FTC), San Francisco, CA, USA, 6–7 December 2016; pp. 731–738. [CrossRef]

15. Tan, L.; Wang, N. Future internet: The Internet of Things. In Proceedings of the 2010 3rd International Conference on Advanced Computer Theory and Engineering (ICACTE), Chengdu, China, 20–22 August 2010; pp. V5-376–V5-380. [CrossRef]

16. Internet of Things, Wikipedia. Available online: https://en.wikipedia.org/w/index.php?title=Internet_of_things&oldid=1096416377 (accessed on 5 July 2022).

17. Jamali, M.A.J.; Bahrami, B.; Heidari, A.; Allahverdizadeh, P.; Norouzi, F. (Eds.) IoT Architecture. In *Towards the Internet of Things: Architectures, Security, and Applications*; Springer International Publishing: Cham, Switzerland, 2020; pp. 9–31. [CrossRef]

18. GDPR. A User-Friendly Guide to General Data Protection Regulation (GDPR). Available online: https://www.gdpreu.org/ (accessed on 5 June 2022).

19. EUR-Lex. EUR-Lex—32016R0679—EN. Available online: https://eur-lex.europa.eu/legal-content/EN/TXT/?uri=CELEX:32016R0679,Mar.30,2021 (accessed on 5 June 2022).

20. Blackmer, W.S. GDPR: Getting Ready for the New EU General Data Protection Regulation. *InfoLawGroup LLP* **2016**, *22*, 2016.

21. General Data Protection Regulation, Wikipedia. Available online: https://en.wikipedia.org/w/index.php?title=General_Data_Protection_Regulation&oldid=1089437849 (accessed on 6 June 2022).

22. Wilhelm, E.O. A Brief History of the General Data Protection Regulation (1981–2016). Available online: https://iapp.org/resources/article/a-brief-history-of-the-general-data-protection-regulation/ (accessed on 6 June 2022).

23. Council of the EU. Data Protection Reform: Council Adopts Position at First Reading. Available online: https://www.consilium.europa.eu/en/press/press-releases/2016/04/08/data-protection-reform-first-reading/ (accessed on 6 June 2022).

24. Regulation of the European Parliament and of the Council on the Protection of Natural Persons with Regard to the Processing of Personal Data and on the Free Movement of Such Data, and Repealing Directive 95/46/EC (General Data Protection Regulation)-First Reading, Adoption of the Council's Position at First Reading. Available online: https://reurl.cc/ZApkLW (accessed on 6 June 2022).

25. EUR-Lex—32016L0680—EN. Available online: https://eur-lex.europa.eu/legal-content/EN/TXT/HTML/?uri=OJ:L:2016:119:FULL (accessed on 6 June 2022).

26. Li, L.; Chen, C. Differences and Reconciliation between GDPR and Taiwan's Individual Capital Laws. Available online: https://view.ctee.com.tw/tax/10989.html (accessed on 1 July 2022).

27. Nakamoto, S. Bitcoin: A Peer-to-Peer Electronic Cash System. Available online: https://bitcoin.org/bitcoin.pdf (accessed on 6 June 2022).

28. Dwork, C.; Lynch, N.; Stockmeyer, L. Consensus in the presence of partial synchrony. *J. ACM* **1988**, *35*, 288–323. [CrossRef]

29. Lamport, L. Paxos Made Simple. *ACM Sigact News* **2001**, *32*, 18–25.

30. Dwork, C.; Naor, M. Pricing via Processing or Combatting Junk Mail. In *Advances in Cryptology—CRYPTO'92*; Lecture Notes in Computer Science; Springer: Berlin, Heidelberg, Germany, 1993; Volume 740, pp. 139–147. [CrossRef]

31. Bentov, I.; Pass, R.; Shi, E. Snow White: Provably Secure Proofs of Stake. Available online: https://ia.cr/2016/919 (accessed on 5 June 2022).

32. Saad, S.M.S.; Radzi, R.Z.R.M. Comparative Review of the Blockchain Consensus Algorithm Between Proof of Stake (POS) and Delegated Proof of Stake (DPOS). *Int. J. Innov. Comput.* **2020**, *10*, 27–32. [CrossRef]

33. Dziembowski, S.; Faust, S.; Kolmogorov, V.; Pietrzak, K. Proofs of Space. In *Advances in Cryptology—CRYPTO 2015*; Springer-Verlag: Berlin, Germany; Heidelberg, Germany, 2015; pp. 585–605. [CrossRef]

34. Howard, H.; Mortier, R. Paxos vs Raft have we reached consensus on distributed consensus? In Proceedings of the 7th Workshop on Principles and Practice of Consistency for Distributed Data, Heraklion, Greece, 27 April 2020; pp. 1–9. [CrossRef]

35. Castro, M.; Liskov, B. Practical Byzantine fault tolerance. In Proceedings of the 3rd Symposium on Operating Systems Design and Implementation (OSDI 1999), New Orleans, LA, USA, 22–25 February 1999; pp. 173–186.

36. Mathieu, B.; Ching, A.; Chursin, A.; Danezis, G.; Garillot, F.; Li, Z.; Malkhi, D.; Naor, O.; Perelman, D.; Sonnino, A. State Machine Replication in the Libra Blockchain. Available online: https://developers.diem.com/papers/diem-move-a-language-with-programmable-resources/2019-06-18.pdf (accessed on 6 June 2022).

37. Pan, J.; Song, Z.; Hao, W. Development in Consensus Protocols: From PoW to PoS to DPoS. In Proceedings of the 2021 2nd International Conference on Computer Communication and Network Security (CCNS), Xining, China, 30 July–1 August 2021; pp. 59–64. [CrossRef]

38. Alfandi, O.; Otoum, S.; Jararweh, Y. Blockchain Solution for IoT-based Critical Infrastructures: Byzantine Fault Tolerance. In Proceedings of the NOMS 2020–2020 IEEE/IFIP Network Operations and Management Symposium, Budapest, Hungary, 20–24 April 2020; pp. 1–4. [CrossRef]
39. Lamport, L.; Shostak, R.; Pease, M. The Byzantine Generals Problem. *ACM Trans. Program. Lang. Syst.* **1982**, *4*, 382–401. [CrossRef]
40. Official Legal Text. General Data Protection Regulation (GDPR). Available online: https://gdpr-info.eu/ (accessed on 13 June 2022).
41. Zaeem, R.N.; Barber, K.S. A study of web privacy policies across industries. *J. Inf. Priv. Secur.* **2017**, *13*, 169–185. [CrossRef]

 *electronics*

Article

The Development of an Autonomous Vehicle Training and Verification System for the Purpose of Teaching Experiments

Chien-Chung Wu *, Yu-Cheng Wu and Yu-Kai Liang

Department of Computer Science and Information Engineering, Southern Taiwan University of Science and Technology, No. 1, Nantai Street, Tainan 71005, Taiwan

* Correspondence: wucc@stust.edu.tw; Tel.: +886-6-2533131 (ext. 3235)

Abstract: To cultivate students' skills in building autonomous vehicle neural network models and to reduce development costs, a system was developed for on-campus training and verification. The system includes (a) autonomous vehicles, (b) test tracks, (c) a data collection and training system, and (d) a test and scoring system. In this system, students can assemble the hardware of the vehicle, configure the software, and choose or modify the neural network model in class. They can then collect the necessary data for the model and train the model. Finally, the system's test and scoring system can be used to test and verify the performance of the autonomous vehicle. The study found that vehicle turning is better controlled by a motor and steering mechanism, and the camera should be mounted in a high position and at the front of the vehicle to avoid interference with the steering mechanism. Additionally, the study revealed that the training and testing speeds of the autonomous vehicle are dependent on each other, and high-quality results cannot be obtained solely by training a model based on camera images.

Keywords: self-driving car; donkey car; autonomous car

Citation: Wu, C.-C.; Wu, Y.-C.; Liang, Y.-K. The Development of an Autonomous Vehicle Training and Verification System for the Purpose of Teaching Experiments. *Electronics* **2023**, *12*, 1874. https://doi.org/10.3390/electronics12081874

Academic Editor: Felipe Jiménez

Received: 20 March 2023
Revised: 6 April 2023
Accepted: 14 April 2023
Published: 15 April 2023

1. Introduction

The frequent occurrence of traffic accidents has allowed for extensive statistical analysis, which shows that more than 90% of traffic accidents are caused by human negligence. Research by Boverie et al. [1] showed that more than 20% of traffic accidents were caused by drivers' hypo-vigilance, such as dozing off or distracted driving. Therefore, governments around the world have been working diligently to try to reduce the occurrence of traffic accidents. Some experts estimate that about 70% of traffic accidents can be avoided with the assistance of automated driving systems or autonomous driving technologies. In order to encourage schools and manufacturers to invest in the research and development of relevant technologies, the United States held the first DARPA Grand Challenge [2] in the Mojave Desert region of the United States in 2004, starting the research and development competition for autonomous driving technologies, which also contributed to the launch of Google's Self-Driving Car and Tesla's autonomous commercial vehicles.

With the rapid development in artificial intelligence and deep learning technologies in recent years, research related to autonomous driving technology through deep learning has also begun to flourish. Navarro et al. [3] proposed using a camera with a deep neural network to develop an automatic driving system and collect information through a camera installed in front of the vehicle for automatic driving. In addition, Bojarski et al. [4–6] designed a system that could input the collected images of the road in front of the vehicle into a neural network, extract image features through multi-layer convolutional layers, perform nonlinear operations on features through multiple sets of dense layers, and then conduct regression to output the value (normalized in the range from 0.0 to 1.0) of the steering angle of the vehicle to achieve automatic driving.

However, considering the cost and the feasibility of a project for teaching experiments, the construction of a large-scale system architecture is not easy to achieve. Therefore, one option to consider is to lower the threshold for learning and verifying autonomous driving technology. A variety of simulation systems and small vehicle verification systems have been developed so far, including (a) AWS DeepRacer [7], (b) Udacity's Autonomous Car Simulator [8], (c) CARLA Simulator [9], and (d) Donkey Car Simulator [10].

AWS DeepRacer is a fully autonomous vehicle simulation and training environment. In this environment, the Amazon corporation uses cloud services to implement reinforcement learning in the cloud. The AWS DeepRacer Evo car is commercially available on amazon.com and is equipped with stereo cameras and an LiDAR sensor. Forward-facing left and right cameras constitute stereo cameras, which allows the vehicle to learn depth information in images. This information can then be used to sense and avoid approaching objects on a track. The LiDAR sensor is backward-facing and detects objects behind and to the sides of the vehicle. From the perspective of education, Cota et al. [11] analyzed and proposed blueprints for the future training of autonomous vehicles through reinforcement learning using the AWS DeepRacer.

Udacity's Autonomous Car Simulator primarily uses training data generated by driving a car to simulate autonomous driving and uses PilotNet [12] to set up a set of virtual cameras on the left, middle, and right sides of the simulator to collect driving images. The software has two different modes: a training mode and an autonomous mode. In training mode, the user can control and drive the vehicle manually through the keyboard or a mouse. In this mode, there are three cameras set up in front of the vehicle to record the driving behavior and the steering angle, driving speed, throttle, and brake data. Technically, the simulator acts as a server from which the program can connect and receive a stream of image frames. With enough driving data collected, the system is trained with the PilotNet neural network. In autonomous mode, the machine learning model can be tested through a network interface. On top of that, this system also provides two different tracks for users to choose from, and it is considered a good entry-level system for autonomous vehicle research. There are no complex roads, no traffic signs, no buildings, no pedestrians, and no vehicles designed as obstacles in the system. Despite being a very basic software package for simulating autonomous driving, Udacity's Autonomous Car Simulator is an open-source project.

CARLA Simulator is a complete simulation environment that can be used for autonomous vehicle testing. It uses Unreal Engine as the basic engine software for field object calculation. In addition to the built-in scenes provided, the system also supports users to define street scenes and interactive scenarios, and supports external sensors or related modules, including RGB Cameras, IMUs (Inertial Measurement Units), RADAR (Radio Azimuth Direction and Ranging), DVSs (Dynamic Vision Sensors), etc. Apart from user-defined sensing modules, this system can also switch between different weather conditions for simulation and testing. It is a simulation system virtually identical to the real environment. It can also support SUMO co-simulation mode, which can be used to randomly simulate multiple vehicles driving on the road at the same time. In the CARLA environment, complex traffic scenarios can be simulated and verified through the simulator. Óscar et al. [13,14] used the ROS framework through deep reinforcement learning to verify autonomous driving applications on the CARLA simulator. Terapaptommakol et al. [15] proposed using a deep Q-network method in the CARLA simulator to develop an autonomous vehicle control system that achieves trajectory design and collision avoidance with obstacles on the road in a virtual environment. This approach allows for the avoidance of collisions with obstacles and enables the creation of optimized trajectories in a simulated environment. In addition, Gutiérrez-Moreno et al. [16] presented an approach to intersection handling in autonomous driving, specifically the use of a deep reinforcement learning approach with curriculum learning and the effectiveness of the Proximal Policy Optimization algorithm in inferring desired behavior based on the behavior of adversarial vehicles in the CARLA simulator.

Donkey Car, similar to the AWS DeepRacer, is a simulation system that provides both a virtual autonomous environment and physical vehicles. Donkey Car is also an open-source project, through which users or manufacturers can develop or modify the software and hardware, and there are many related products on the market that can be purchased. However, the primary problem with this system is that although the training methods of the real and simulated environments are the same, the results of training a neural network using data collected on real-world roads cannot be directly applied to enable a vehicle to drive on simulated roads in a simulator. Similarly, the results of training a neural network using data collected in a simulator cannot be directly applied to enable a real-world vehicle to drive on real-world roads. In simple terms, the simulation software and the physical vehicles belong to two different training settings. Only when the training track is exactly the same as the test track can the autonomous vehicle be verified on the same track in the simulation system after the software has been trained. Overall, however, Donkey Car is an excellent system for autonomous vehicle research, with an open-source autonomous platform providing all the necessary details for users, and there are numerous ways to use the simulator, depending on one's goals.

2. System Design

Based on Donkey Car, the autonomous driving training and verification system in this paper was designed to be used in teaching experiments. In addition to setting up autonomous vehicles freely, developers can adjust the model in the environment for training. At the same time, developers can objectively evaluate the advantages and disadvantages of the model on the automatic test and scoring system. The design of this system is explained as follows.

The system consists of four components: (a) autonomous vehicles, (b) test tracks, (c) the data collecting and training system, and (d) the test and scoring system. The autonomous vehicles use Raspberry Pi as the core and are installed with a Raspberry PI OS and the Donkey Car environment for management. Users design the geometric shapes of the test tracks, which are combined through poster printing. The data collection and training system is a computer running Ubuntu OS with the Donkey Car environment installed. Its main function is to receive data collected by the autonomous vehicles, perform post-data processing and training, and return the results to the autonomous vehicles. The test and scoring system is a computer with Microsoft Windows 11, and installed with Anaconda, PyCharm community, Yolo V7 [17], and the software proposed in this paper. It processes real-time images captured by the overhead camera of the test track and determines if the wheels of the autonomous vehicles touch the track or go beyond the boundary. The system block diagram is shown in Figure 1.

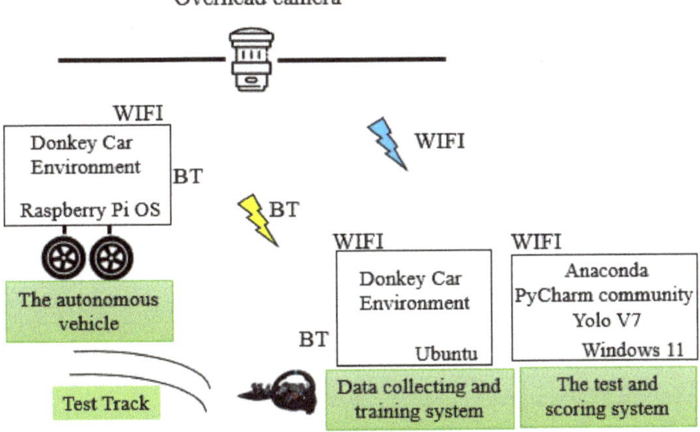

Figure 1. System block diagram.

2.1. Autonomous Vehicles

There were two kinds of design for autonomous vehicles in this paper, which we will designate Vehicle A and Vehicle B. Both used a Raspberry Pi 4 B, which is manufactured by the Raspberry Pi Foundation headquartered in London, United Kingdom, as the hardware core for the computing system, Donkey Car version 4.2.1 as the software, the DC Gearbox Motor as the vehicle drive motor, and the L298N as the motor controller. The primary difference between the two was the steering mechanism. For the steering mechanism, Vehicle A was steered by controlling the rotational speed difference between the left and right wheels, as shown in Figure 2a. For Vehicle B, the front wheel steering mechanism was achieved by controlling the steering of the servo motor, as shown in Figure 2b. The hardware block diagrams of the two vehicles are shown in Figure 2.

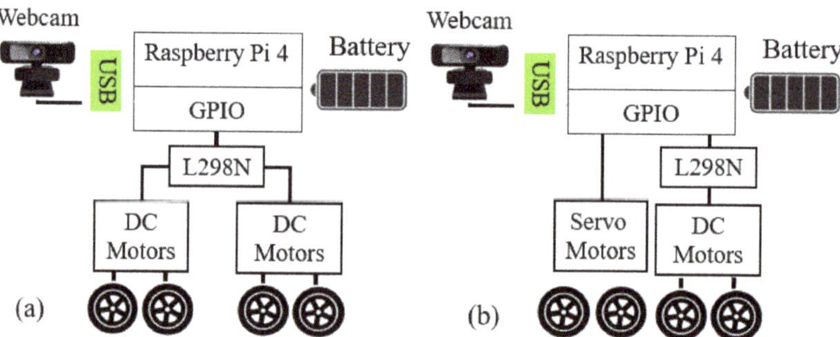

Figure 2. Hardware block diagrams of two vehicles, (**a**) Vehicle A is steered by controlling rotational speed differences between the left and right wheels, while (**b**) Vehicle B is steered by controlling the servo motor.

2.2. Test Track

The test track is assembled by piecing together output posters, and users can adjust or design the geometry of the track during testing. This design enables the testing method of the system to be more flexible. Track A is a closed ellipse track with an S-shaped curve, as shown in Figure 3a. Track B is a closed ellipse track, shown in Figure 3b.

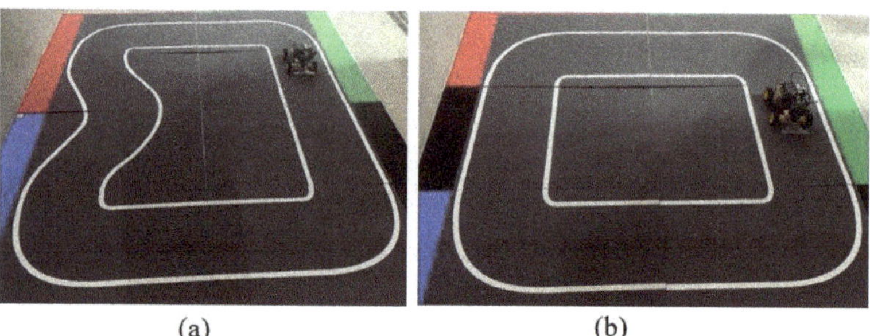

Figure 3. Vehicle test tracks, (**a**) Track A, a closed ellipse track with an S-shaped curve, and (**b**) Track B, a closed ellipse track.

2.3. Data Collection and Training System

The entire autonomous vehicle training and verification process is shown in Figure 4. The trainer supervises the overall situation or the front-facing image of the remote vehicle through visual inspection or a computer screen, respectively, and operates the Xbox racing wheel and pedal to control the steering of the vehicle, as shown in Figure 4a. The speed of the vehicle was controlled by pedaling the accelerator or the brake, and the driving data were collected. When the driving data collection was complete, redundant and blurred images were corrected or removed, as shown in Figure 4b, and then the neural network was trained by collecting and correcting images and data in the direction of vehicle travel. A screenshot of the training screen is shown in Figure 4c.

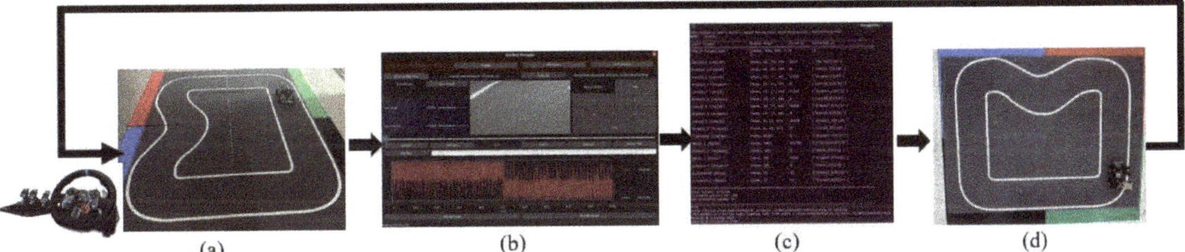

Figure 4. Flow of the autonomous vehicle training and verification process, (**a**) driving for data collection, (**b**) data correction, (**c**) training through the neural network, and (**d**) testing and scoring.

In the test mode, when the training of the neural network was complete, the trained weight files and neural network description files were transplanted into the autonomous vehicle. The autonomous vehicle predicted the steering angle of the vehicle through the neural network based on the front-facing images that were captured. The system converted the angle into corresponding parameters and sent them to the servo motor, which controlled the steering mechanism through GPIO, which allowed the vehicle to drive by itself on the track. The experimenters conducted scoring and verification through the test and scoring system, as shown in Figure 4d. The screenshots of the flow of the autonomous vehicle training and verification process are shown in Figure 4.

2.4. Testing and Scoring System

In order to evaluate the performance of the autonomous vehicle on the test track, a test and scoring system for the autonomous vehicle was set up. The scoring of the autonomous vehicle was achieved by first setting up an overhead camera, with the camera facing downward towards the track, and then using machine vision technology to conduct image processing and recognition. The system needed to first find the track, then locate the position of the vehicle, and after collecting images of the wheels for neural network training, it was able to identify the exact position of the wheels. Finally, it is possible to determine if a vehicle has gone off the track by checking if the coordinates of the vehicle's wheels intersect with the lane markings. The identification and processing methods are shown in Figure 5.

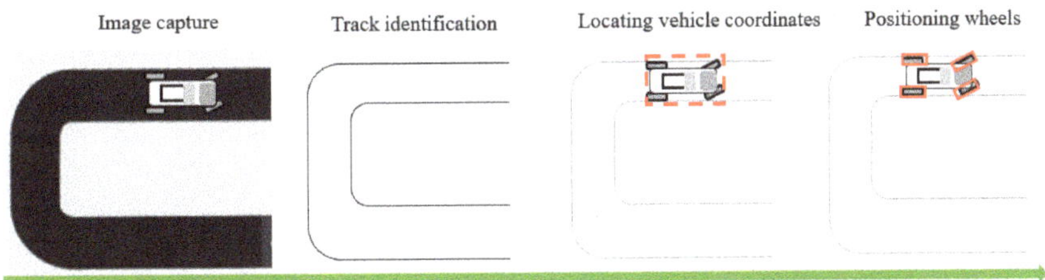

Figure 5. Schematic diagram of processing method for vehicle off-track detection.

2.4.1. Equipment Installation

In order to recognize whether the vehicle was driving on the track, there was an overhead camera set up in the system, and the camera was installed facing the track, as shown in Figure 6.

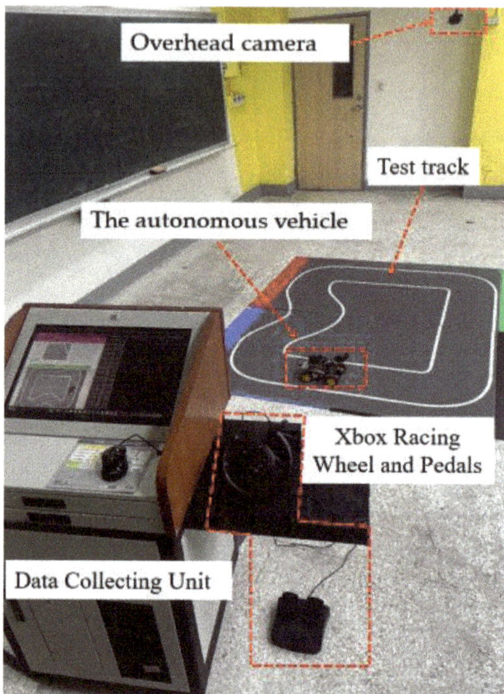

Figure 6. Photo of the system installation.

2.4.2. Explanation of the Testing and Scoring System

When system execution began, the system first captured an image with no footage of a vehicle as the test background, as shown in Figure 7a. The image was converted into grayscale and denoised; then, the Canny edge detection algorithm was used to find the edge features of the track. Finally, the findContours function was used to detect the contour for the extraction of the lane markers of the track, as shown in Figure 7b.

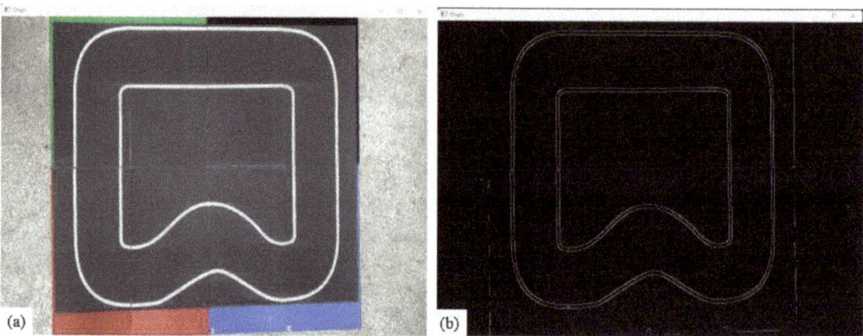

Figure 7. Marking the lane markers of the test track, (**a**) the image of the test track background, and (**b**) the image of the lane marker of the test track.

When the system started scoring, the system sequentially processed each frame of images captured by the overhead camera. The vehicle position was obtained by subtracting the background image from the acquired image. The schematic diagram of positioning vehicle coordinates is shown in Figure 8.

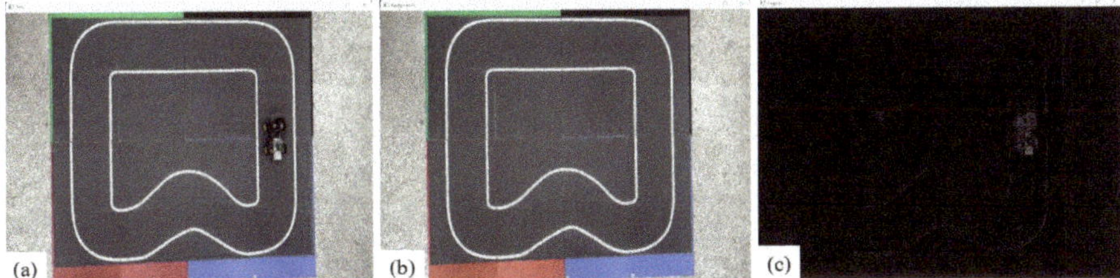

Figure 8. Schematic diagram of the program positioning vehicle coordinates: (**a**) image of a vehicle driving on the track, (**b**) test track background, and (**c**) image of vehicle positioning.

As the overhead camera was installed pointing at the track, when the autonomous vehicle was driving, this system could identify whether the wheels of the vehicle touched the lane marker or not and for how long it went off the track. As an example, in the orthographic projection of the vehicle body exhibited in Figure 9 below, the vehicle crosses the track line, but its wheels do not actually touch the lane marker, as shown in Figure 9a. Conversely, the system must be able to detect correctly if a wheel even only slightly touches a lane marker, as shown in Figure 9b. Therefore, instead of scoring lane integrity based on the outline of the vehicle body, it was decided that the system needed to correctly locate the outline of the wheels.

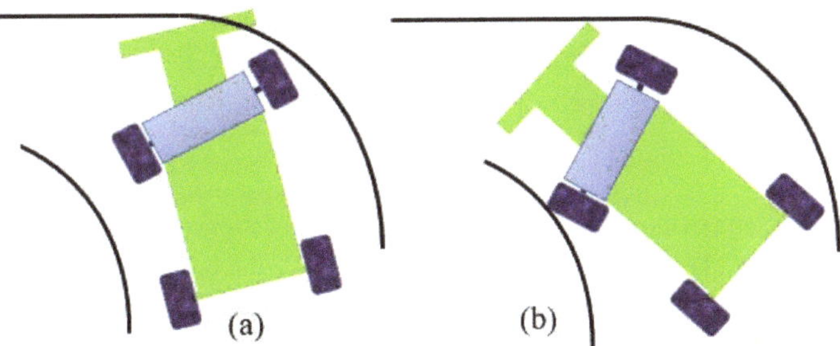

Figure 9. Schematic diagram of lane-touching conditions of the autonomous vehicle, (**a**) vehicle body crossing the track line not counting as off-track, and (**b**) wheel touching the lane marker counting as off-track.

In order to correctly locate the wheel coordinates, Yolo v7 was used as the neural network for recognizing wheels in this study. While the vehicle was driving under different rotation angles, images were simultaneously collected by the overhead camera. After subtracting the background image from the acquired images, the four wheels of the vehicle in the images were labeled and then trained through the Yolo v7 neural network. Finally, a system that could correctly recognize and locate the wheels was developed. The training process of the wheel image recognition system for autonomous vehicles is shown in Figure 10.

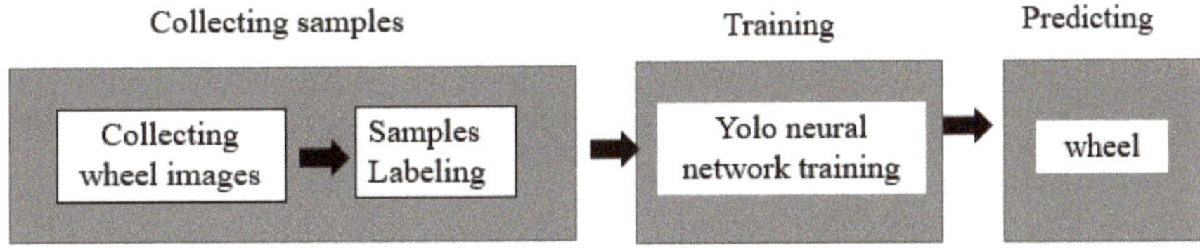

Figure 10. The training process of wheel image recognition system for autonomous vehicles.

The position of the relevant object as recognized by Yolo v7 was marked with a rectangular box. The wheels, however, drove at various angles while the vehicle was in the process of turning.

Figure 11a is a schematic diagram exhibiting the wheels' angles during vehicle turning. The red boxes frame the wheels as recognized through Yolo v7; this is schematized in Figure 11b. The red rectangles here clearly do not precisely represent the outline of the wheels. Therefore, an algorithm was developed to obtain the actual wheel contours by performing an intersection operation between the vehicle image obtained and the red rectangle regions extracted by Yolo V7; this is schematized in Figure 11c.

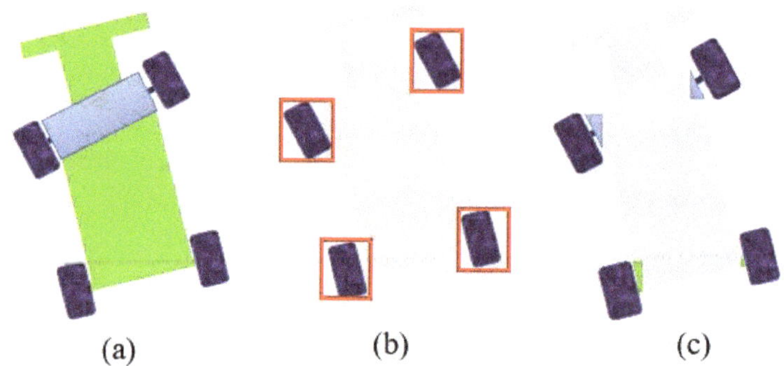

Figure 11. Schematic diagram of the turning vehicle,:(**a**) a top-down view of the vehicle, (**b**) locating the wheels, and (**c**) obtaining the wheels' outline.

2.4.3. Testing and Scoring Algorithm

In this paper, the background image without vehicles is defined as $Frame_{Nocar}$ and the image with the vehicle is defined as $Frame_{car}$. $Frame_{Nocar}$ was subtracted from $Frame_{car}$ through Formula (1) to obtain the vehicle image $Image_{Car}$, and then the pre-trained Yolo v7 was used to locate the four wheels, defined as $Wheels_{Yolo}$. Using Formula (2) to determine the overlapping area of the $Image_{Car}$ and $Wheels_{Yolo}$ images, the final range of the $Wheels_{Edge}$ is produced. In the following formulas, adding '(t)' to the formula means that the frame read at time t.

$$Image_{Car}(t) = Frame_{car}(t) - Frame_{Nocar} \tag{1}$$

$$Wheels_{Edge}(t) = Image_{Car}(t) \cap Wheels_{Yolo}(t) \tag{2}$$

All that is required is to ascertain whether there was an overlap between the outlines of the wheels and the track lane marker to judge that the driving vehicle had gone off the track. $Frame_{Nocar}$ was computed through the Canny edge detection algorithm to find the edge features of the track, and the findContours function was used to define the coordinates sets of the contour as $Track_{Contours}$, as shown in Formula (3).

$$Track_{Contours}(t) = findContours(CannyEdgeDetection(Frame_{Nocar}(t))) \tag{3}$$

Next, $Wheels_{Edge}$ was computed through the Canny edge detection algorithm to find the edge features of the vehicle wheels, and the findContours function was used to define the coordinate sets of the contour as $Wheels_{Contours}$, as shown in Formula (4).

$$Wheels_{Contours}(t) = findContours(CannyEdgeDetection(Wheels_{Edge}(t))) \tag{4}$$

Finally, an intersection function was used to determine whether there was an overlap between $\text{Track}_{\text{Contours}}$ and $\text{Wheels}_{\text{Contours}}$ to determine if the vehicle had driven onto the track lane marker $\text{Output}_{\text{OutofBounds}}$, as in Formula (5).

$$\text{Output}_{\text{OutofBounds}}(t) = \text{intersection}\ (\text{Wheels}_{\text{Contours}}(t),\ \text{Track}_{\text{Contours}}(t)) \qquad (5)$$

When the test and scoring system was executed, two values were generated: the first being the number of times that the wheels of the autonomous vehicle touched the track lane marker and the second being the number of frames in which the wheels of the autonomous vehicle touched the track lane marker, defined as follows.

The number of times the wheels of the autonomous vehicle touched the lane marker of the track was defined as $\text{Counter}_{\text{touch}}$. Once the wheels of the vehicle in the captured image touched the lane marker, going off and then back on to the track again, the count of $\text{Counter}_{\text{touch}}$ would be increased by 1, as in Formula (6), primarily to keep track of the number of times the autonomous vehicle had touched the track lane marker. The symbol $\varnothing$ represents an empty set and n represents a positive integer.

$$\text{Counter}_{\text{touch}} = \text{Counter}_{\text{touch}} + 1,$$
$$\text{if } (\text{Output}_{\text{OutofBounds}}(t) \neq \varnothing \text{ and } \text{Output}_{\text{OutofBounds}}(t+n) = \varnothing \text{ and } n > 0) \qquad (6)$$

However, there might be an omission of relevant data if one evaluates an autonomous vehicle based only on the number of times the wheels touch a lane marker. One needs to consider if the wheels immediately correct back to the track after touching a lane marker, or if they go off the track for a significantly longer time before correcting back, which constitute two different conditions. In order to be able to distinguish between these two conditions, $\text{Counter}_{\text{touchFrame}}$ was specifically defined in this paper to represent each frame in the video captured by the system that recognized wheels touching a lane marker, which increases the count of $\text{Counter}_{\text{touchFrame}}$ by 1, as in Formula (7).

$$\text{Counter}_{\text{touchFrame}} = \text{Counter}_{\text{touchFrame}} + 1, \text{ if } \text{Output}_{\text{OutofBounds}}(t) \neq \varnothing \qquad (7)$$

During testing, if any of the wheels of the autonomous vehicle touched the lane marker of the track and failed to correct back to the track, causing the vehicle to run off the track, the test would be judged as a failure.

Figure 12 is a screenshot of the execution of the test and scoring system. It shows the position of the driving vehicle and the four wheels. The yellow number in the upper left corner indicates the number of times the vehicle touched the lane marker, $\text{Counter}_{\text{touch}}$, and the red number in the upper left corner indicates the number of video frames in which the vehicle touched the lane marker, $\text{Counter}_{\text{touchFrame}}$, during the test.

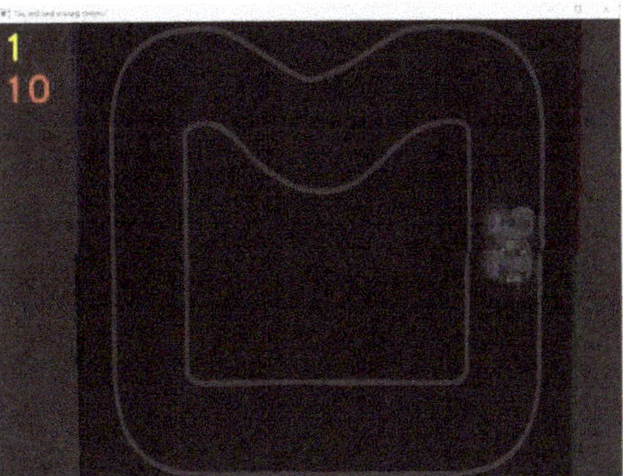

Figure 12. Screenshot of the execution of the test and scoring system.

2.5. The Process of the Course

Figure 13 is a flowchart depicting the process of course execution. In this course, students learned how to assemble an autonomous car, install and configure the necessary software, and perform a series of tests to ensure that the car was functioning properly. In the interests of reproducibility, we relate the concrete steps that the students covered:

Step 1: Students assembled the car hardware, including the chassis, motors, motor drivers, battery pack, wheels, Raspberry Pi, webcam, and wiring.

Step 2: Students installed the Raspberry Pi and set up the Donkey Car environment, including configuring the necessary hardware drivers. For example, they needed to add a driver for their hardware chip to the actuator.py file in the Donkey Car project's parts folder. Then, in the manage.py and myconfig.py files, they needed to specify the hardware driver they just set up.

Step 3: Students tested the functionality of the Donkey Car. They started by lifting the assembled car off the ground and running the command "python manage.py drive". This started the car and allowed them to verify that the webcam was streaming properly and that the mechanism and wheels were turning correctly.

Step 4: Students used two different methods provided by the system to operate the vehicle and collect training data. The first method involved manually driving the vehicle and collecting data by entering the command "python manage.py drive". Prior to starting, they could access a menu by entering "http://vehicle-IP:8887" into a web browser on a computer or mobile device, where"vehicle-IP" refers to the Wi-Fi IP of the autonomous car. In the menu, they could set the value for "Max Throttle" and select "User(d)" in the "Mode and Pilot" menu to configure the parameters for remote control of the vehicle.

The second method involved using an Xbox racing wheel and pedals to control the vehicle via Bluetooth by entering the command "python manage.py drive –js". The direction of the vehicle could be controlled using the steering wheel, while the throttle could be adjusted using the pedals to increase or decrease the throttle value.

Step 5: Students uploaded the collected data to a server or computer for further processing.

(Note: Steps 6 through 9 were all carried out on a server or computer.)

Step 6: Students modified the Donkey Car neural network model on their computer.

Step 7: Driving the car may cause it to leave the track boundary or collect too much repetitive data; thus, before training the neural network, students needed to delete or correct any data that included improper driving or unclear images.

Step 8: Students trained the model by running the "python train.py –tub <path_to_collected_data> –model <name_of_output_model_file>" command.

Step 9: Students downloaded the completed weight file from the trained model to the autonomous car for testing.

Step 10: Students launched the testing program with "python manage.py drive –model trained_Model" and configured the maximum throttle value and mode as well as pilot menu setting "Local Pilot(d)" in the browser window at "http://vehicle-IP:8887". Then, they tested the car using the testing and scoring system. The system measured the car's performance, including its ability to stay on the track and avoid obstacles.

Step 11: If the test results did not meet requirements, they went back to step 4 to collect the data and train the model again.

By the end of the process, students learned how to assemble and configure an autonomous car, collect training data, and train and test a neural network model for autonomous driving.

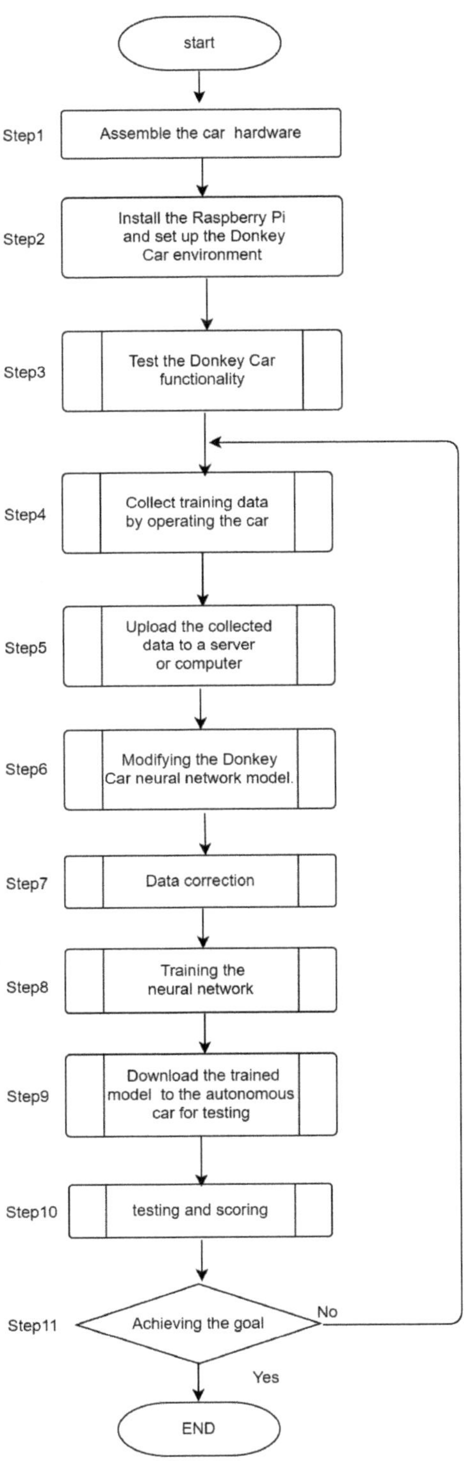

Figure 13. A flow chart depicting the process of course execution.

3. Experiment and Results

Before using the test and scoring system, this system must undergo verification. The following describes the verification process for this system.

3.1. Verification of the Test and Scoring System

As the first stage of the experiment, testing and verification had to be conducted for the test and scoring system. The operator used throttle of 20%, 35%, and 50% to remotely control the autonomous vehicle driving on the test track using the Xbox racing wheel and pedal. During the process, the operator intentionally made the vehicle's wheels touch the lane marker at different locations, and through this process simulated the driving path of the vehicle; this was recorded with the overhead camera.

The next step was to capture frames from the three videos recorded under different throttle conditions, have personnel check them one by one, and manually record the results of $Counter_{touch}$ and $Counter_{touchFrame}$. Meanwhile, the three different videos were tested separately through the test and scoring system. It was thereby confirmed that the results obtained through system detection were consistent with the results obtained from personnel checking. With the verification of the test and scoring system completed, the evaluation of the performance of the autonomous vehicles could commence.

3.2. Experimental Conditions

The default training model for the Donkey Car, Keras Linear, was used to train all the autonomous vehicles in this study. The track featured an S-curve, turns, and straight sections. During training, the car was operated by personnel, who found it difficult to operate correctly with a throttle value greater than 50% on this short and varied track. On the lower end, a throttle value set to less than 20% tended to cause the car to get stuck at S-curve turns. Therefore, for this experiment, the throttle value was set between 20% and 50%, and testing was conducted using a mid-range value of 35%. The test procedure involved running autonomous Vehicles A and B on both Track A and Track B for 10 laps each and recording the results. However, after confirming that Vehicle B outperformed Vehicle A in steering performance, only Vehicle B was used in subsequent experiments.

The battery of the autonomous vehicle was fully charged for each experiment. In this system, 100% throttle represented full power, with 50% throttle corresponding to half power used to control the accelerator. To collect training data, both vehicles were driven with a throttle value of 35% for 20 laps each on Track A.

All data collection, training, and testing procedures for the autonomous vehicles followed the steps described in Section 2.5 above from Step 4 to Step 11.

In the preliminary stage of the study, it was found that the autonomous vehicle did not operate accurately on the track, and it was observed that three primary problems needed to be resolved: (a) the impact of vehicle steering mechanism on the test results of the autonomous vehicles, (b) the impact of the location of the camera on the test results of the autonomous vehicles, and (c) the correlation between the throttle for the training vehicle and the throttle for the testing vehicle.

Next, we conducted tests to come up with the appropriate solutions for these three problems.

3.3. Impact of the Vehicle Steering Mechanism on the Test Results of Autonomous Vehicles

In order to find out whether different steering mechanisms impact the training and verification of the autonomous vehicle, two vehicles with different steering mechanisms were designed for this study.

Vehicle A was steered by controlling the rotational speed difference between the left and right wheels, as shown in Figure 14a, and Vehicle B was steered by controlling a servo motor, as shown in Figure 14b. The photos of two vehicles are shown in Figure 14.

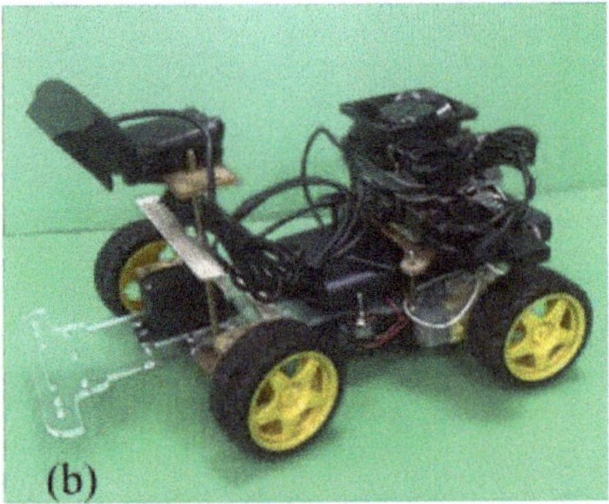

Figure 14. Photos of the two vehicles: (**a**) Vehicle A, steered by controlling the rotational speed difference between left and right wheels, and (**b**) Vehicle B, steered by controlling a servo motor.

The camera setup in the two vehicles was installed at the same position. There were two types of tracks in the experimental design, as shown in Figure 3a,b; they are designated Track A and Track B, respectively.

The two vehicles used the same neural network for training in order to facilitate a comparison of the effects of different steering mechanisms on vehicle autonomous training. The experimental conditions are as described in Section 3.2, "Experimental Conditions".

As shown in Table 1, when testing on Track A, the wheels of Vehicle A touched the lane marker more frequently while driving on an S-shaped curve. The test results showed that the number of times the wheels of Vehicle A touched the lane marker was 32, and the number of video frames in which the wheels touched the lane marker was 526. The number of times the wheels of Vehicle B touched the lane marker was 9, and the number of video frames in which they touched the lane marker was 151. When driving on a closed ellipse Track B, the number of times the wheels of Vehicle A touched the lane marker was 14, and the number of video frames in which the wheels touched the lane marker was 245, whereas the number of times the wheels of Vehicle B touched the lane marker was 4, and the number of video frames that they touched the lane marker was 63.

Table 1. Test results of vehicles driving with 35% throttle with different steering mechanisms on different tracks for 10 laps.

Test Vehicle	Test Track (Results)	Track A		Track B	
		$Counter_{touch}$	$Counter_{touchFrame}$	$Counter_{touch}$	$Counter_{touchFrame}$
Vehicle A		32	526	14	245
Vehicle B		9	151	4	63

As in Table 1, the experimental results showed that Vehicle B, whose steering mechanism was controlled by a servo motor, performed much better than Vehicle A regarding controlling the direction of the vehicle, especially when driving on the S-shaped curve or sharp turns.

3.4. Impact of the Location of the Camera on the Test Results of the Autonomous Vehicles

Since the testing described above found that the design of the steering mechanism had a better effect on controlling the direction of the vehicle, the following test was based on this design. This test addressed whether the camera should be installed higher or lower, and whether it should be installed in front of the vehicle or linked up to the steering mechanism.

In order to determine the best position to install the camera, the camera was set up in three different positions, as shown in Figure 15. Position (I): the camera was installed in front of the vehicle body and set up at a low position. Position (II): the camera was installed in front of the vehicle body and set up at a high position, with it facing downward towards the track. Position (III): the camera was mounted on the vehicle steering mechanism. The direction of the camera view installed at position (I) and position (II) was consistent with the vehicle's travel direction, and the direction of the camera view at position (III) was synchronized with the vehicle's steering direction (that is, not in the vehicle's travel direction). Schematic diagrams of the installation positions are shown in Figure 15.

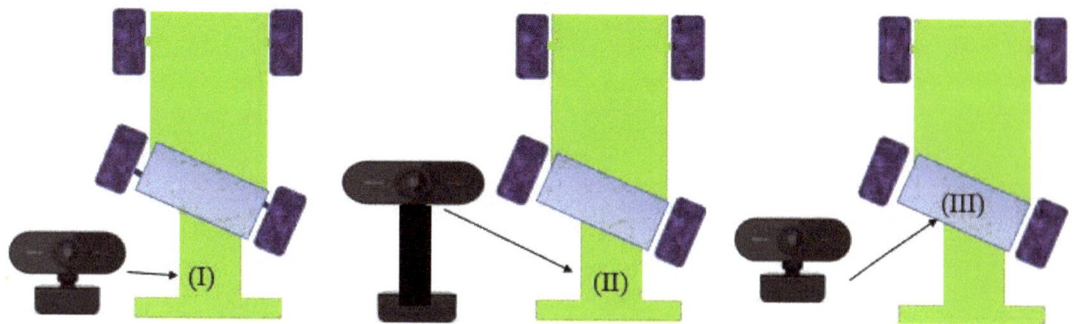

Figure 15. Camera installation positions: low position (I), high position (II), and position linked with the steering mechanism (III).

The procedure for collecting training data was to set up the camera on the vehicle in the three different positions mentioned above, and make the vehicle drive on the track, as shown in Figure 3. With the camera installed at different heights, different views would be presented: the camera installed at a low position in front of the vehicle body (I), which is the view shown in Figure 16a; and the camera at a high position in front of the vehicle body (II), which is the view shown in Figure 16b.

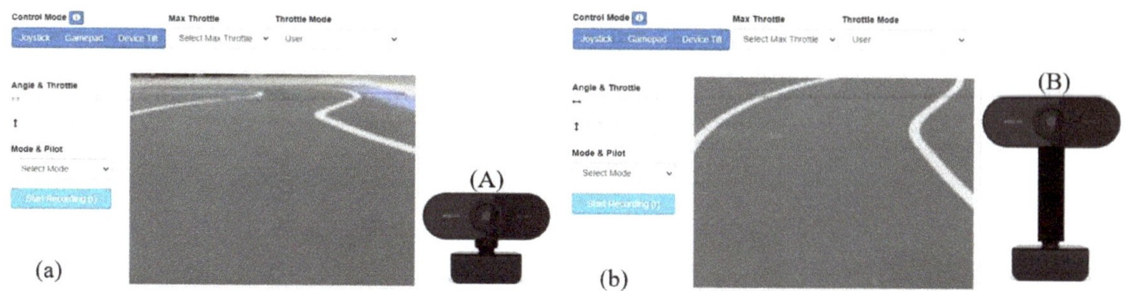

Figure 16. Experimental field: (**a**) the view from position (I), (**b**) the view from position (II).

The experimental conditions are as described in Section 3.2, "Experimental Conditions".

As shown in Table 2, when testing on Track A with the camera set at position (I), the test results showed that the number of times the wheels touched the lane marker was 22, and the number of video frames in which the wheels touched the lane marker was 358. When the camera was set at position (II), the number of times the wheels touched the lane marker was 9, and the number of video frames in which the wheels touched the lane marker was 151. Unexpectedly, the test results showed the outcome was a failure when the camera set at position (III).

Table 2. Comparison results of test driving for ten laps at three camera installation positions.

Camera Installed Position	Test Track (Results)	Track A		Track B	
		$Counter_{touch}$	$Counter_{touchFrame}$	$Counter_{touch}$	$Counter_{touchFrame}$
Position (I)		22	358	11	176
Position (II)		9	151	4	63
Position (III)		fail	fail	fail	fail

As shown in Table 2, when testing on Track B, with the camera set at position (I), the test results showed that the number of times the wheels touched the lane marker was 11, and the number of video frames in which the wheels touched the lane marker was 176. When the camera was set at position (II), the number of times the wheels touched the lane marker was 4, and the number of video frames in which the wheels touched the lane marker was 63. Once again, the test results showed the outcome was a failure when the camera was set at position (III).

The experimental results in Table 2 show that when the camera was set at position (III), most of the captured images were blurred due to the shaking of the camera when turning, and the vehicle would then go off the track during the test, which was thus rated as "failed". When the camera was set at position (I) and position (II), automatic driving was able to be successfully completed. However, it was clear that when the camera was installed in position (II), the effect was much better than it being installed in position (I). The reason for this is that when the camera was installed at position (I), its field of vision facing the track was relatively narrow, and it was prone to falsely detect objects outside the track, which resulted in mistakenly including them as part of the learning data for training, as shown in Figure 16a. When the camera was installed at position (II), however, it was elevated and facing the track, so that the autonomous vehicle could detect the situations ahead of the track much earlier, as shown in Figure 16b, especially where there was a curve ahead.

3.5. Correlation between Throttle for the Training Vehicle and Throttle for the Testing Vehicle

In order to determine the correlation between the throttle for the training vehicle and the throttle for the testing vehicle, tests were carried out as follows: the operator controlled the autonomous vehicles remotely through the Xbox racing wheel and pedal to drive Vehicle A on Track A for 20 laps using successive throttle percentages of 20%, 35%, and 50% to collect data. After training with the collected data, testing was carried of the autonomous vehicle in autonomous mode on Track A for 10 laps at a throttle percentage of 20%, 10 laps at a throttle percentage of 35%, and 10 laps at a throttle percentage of 50%, separately, and the results were recorded using the test and scoring system.

3.5.1. Training and Testing with the Same Throttle

As shown in Table 3, the experimental results show that when using a throttle of 20% for training and testing, the number of times the wheels touched the lane marker was 8, and the number of video frames in which the wheels touched the lane marker was 136. When using a throttle of 35% for training and testing, the number of times the wheels touched the lane marker was 9, and the number of video frames in which the wheels touched the lane marker was 151. Finally, the results showed that when using a throttle of 50% for training and testing, the number of times the wheels touched the lane marker was 9 and the number of video frames in which the wheels touched the lane marker 143.

Table 3. Comparison results of training and testing of the autonomous vehicle for 10 laps with different throttles.

Throttle in Training	Throttle in Testing (Results)	20%		35%		50%	
		Counter_{touch}	$\text{Counter}_{touchFrame}$	Counter_{touch}	$\text{Counter}_{touchFrame}$	Counter_{touch}	$\text{Counter}_{touchFrame}$
20%		8	136	17	274	fail	fail
35%		13	213	9	151	fail	fail
50%		13	217	11	182	9	143

3.5.2. Training and Testing with Different Throttle

As shown in Table 3, the experimental results showed that when the system was trained with 20% throttle and then the autonomous vehicle was tested at 35% throttle, the number of times the wheels touched the lane marker was 17 and the number of video frames in which the wheels touched the lane marker was 274. The experimental results also showed that when the system was trained with 35% throttle and then the autonomous vehicle was tested at 20% throttle, the number of times the wheels touched the lane marker was 13 and the number of video frames in which the wheels touched the lane marker was 213. Finally, the results showed that when the system was trained with 50% throttle and the autonomous vehicle was tested at 20% throttle, the number of times the wheels touched the lane marker was 13 and the number of video frames in which the wheels touched the lane marker was 217. The experimental results showed that the system was trained with 50% throttle and the autonomous vehicle was tested at 35% throttle, the number of times the wheels touched the lane marker was 11 and the number of video frames in which the wheels touched the lane marker was 182.

To summarize, the results were different when the throttle used for training and for test were not the same. We note first that when samples were collected and trained with 20% throttle and then the autonomous vehicle was tested at 35% throttle, the autonomous vehicle completed driving with the wheels occasionally touching the lane marker. When the throttle was increased to 50% for the test, the autonomous vehicle went off the track and failed the test. Second, when samples were collected and trained with 35% throttle and then the vehicle was tested with 20% throttle, the autonomous vehicle functioned normally, with the wheels occasionally touching the lane marker. However, when testing with 50% throttle, the automatic driving could not be completed. Third, when the samples were collected and trained with 50% throttle and then the vehicle was tested at 20% and 35%, the autonomous vehicle functioned normally, with the wheels occasionally touching the lane marker.

3.6. Comparison Results

As shown in Table 4, AWS DeepRacer, Udacity's Autonomous Car Simulator, CARLA Simulator, Donkey Car, and the vehicles designed in this paper were compared. All of them have simulation capabilities, but only AWS DeepRacer, Donkey Car, and the vehicles in this paper have physical cars that can be verified.

The DeepRacer and DeepRacer Evo are priced at USD 399 and USD 598, respectively, while Donkey Car is priced at approximately USD 325. The materials for Vehicle A and Vehicle B, which can be assembled by oneself, cost approximately USD 110 and USD 132, respectively. For AWS DeepRacer, training and model evaluation in the cloud cost USD 3.5/2 h each, and additional storage rental is required. Udacity's Autonomous Car Simulator, CARLA Simulator, and Donkey Car can be installed on a personal computer, so there is no need to pay for usage. Vehicle A and Vehicle B in this paper can be designed to run on self-made tracks, which can be printed on A0 posters for USD 15 per sheet. It is worth noting that this paper introduces a unique test and scoring system, which allows for immediate measurement of the autonomous car's performance after training, specifically with regard to the duration that its wheels are in contact with the road dividers. Overall, the design in this paper has the advantages of being more cost-effective and flexible in terms of increasing or modifying vehicle functions, and provides a more clear and structured way to evaluate and verify performance in testing, especially in educational settings.

Table 4. Comparison results of the items for different methods.

Types \ Items	Physical Vehicles	Prices	Test Environment	Auto Scoring and Verification System
AWS DeepRacer	AWS DeepRacer and Evo car	DeepRacer USD 399 DeepRacer Evo USD 598	AWS Cloud	N/A
Udacity's Autonomous Car Simulator	N/A	N/A	Simulator based on the Unity	N/A
CARLA Simulator	N/A	N/A	Simulator based on the Unreal Engine	N/A
Donkey Car	Donkey Car	USD 325	Simulator based on the OpenAI gym wrapper	N/A
Vehicle A in this paper	Vehicle A	USD 110	Piecing together output posters	Available
Vehicle B in this paper	Vehicle B	USD 132	Piecing together output posters	Available

4. Discussion

We conclude from the above experiments that Vehicle A, with the mechanism of steering through rotational speed difference, performed poorly in controlling the direction of the vehicle, especially when driving on an S-shaped curve or sharp turns.

We also found that when the camera was installed in position (II), the track's front image could be collected much earlier for a turn because of the camera's higher position, and that under the same training mode, a camera installed on a tall vehicle body would have a much better effect on the front images collected for training by the autonomous vehicle.

The above experiments further showed that with a fixed throttle, tests failed under different conditions when only inputting front driving images to control the steering wheel of the autonomous vehicle, due to the difference between the throttle used for the training and the throttle used for the test. Therefore, throttle and front images must be taken into consideration together.

The vehicle design in this study is relatively cheap compared to other platforms, allowing students must have their own autonomous car for implementation in group projects. Although the hardware for the test and scoring system is more expensive, it can be shared during the course for cost efficiency. In addition, the testing track for the vehicles in this study can be customized by combining output posters, allowing for flexibility in adapting to different classroom sizes. Furthermore, the unique design of the auto scoring and verification system in this paper allows for a more concrete and efficient evaluation of the performance of the trained autonomous vehicles.

5. Conclusions

The system designed in this paper describes a complete environment that can be applied in training and verification of autonomous vehicles in schools or research units. Users can modify or replace the Keras Linear model originally used in Donkey Car and can also quickly verify results through the test and scoring system to optimize the neural network. Moreover, through the output of different posters, diverse types of tracks can be assembled freely, such as intersections, multi-directional tracks, etc. This system can provide a teaching and verification platform for schools to conduct autonomous-vehicle-related research courses.

Author Contributions: C.-C.W.: problem conceptualization, methodology, data analysis, writing—review and editing of final draft, results tabulation, and graphic presentation. Y.-C.W.: software development and execution. Y.-K.L.: data collection, training, and execution. All authors have read and agreed to the published version of the manuscript.

Funding: This research was funded by the Ministry of Science and Technology (MOST), Taiwan, grant no. MOST 110-2221-E-218-019.

Data Availability Statement: Not applicable.

Acknowledgments: The authors wish to express their gratitude to the Ministry of Science and Technology (MOST) for supporting this research.

Conflicts of Interest: The authors declare no conflict of interest.

References

1. Boverie, S.; Daurenjou, D.; Esteve, D.; Poulard, H.; Thomas, J. Driver Vigilance Monitoring—New Developments. In Proceedings of the 15th IFAC World Congress on Automatic Control, Barcelona, Spain, 21 July 2002.
2. Behringer, R.; Sundareswaran, S.; Gregory, B.; Elsley, R.; Addison, R.; Guthmiller, W.; Daily, R.; Bevly, D. The DARPA grand challenge—Development of an autonomous vehicle. In Proceedings of the IEEE Intelligent Vehicles Symposium, Parma, Italy, 14–17 June 2004.
3. Navarro, A.; Joerdening, J.; Khalil, R.; Brown, A.; Asher, Z. *Development of an Autonomous Vehicle Control Strategy Using a Single Camera and Deep Neural Networks*; SAE Technical Paper 2018-01-0035; SAE International: Warrendale, PA, USA, 2018. [CrossRef]
4. Bojarski, M.; Testa, D.D.; Dworakowski, D.; Firner, B.; Flepp, B.; Goyal, P.; Jackel, L.D.; Monfort, M.; Muller, U.; Zhang, J.; et al. End to end learning for autonomous cars. *arXiv* **2016**, arXiv:1604.07316v1[cs.CV].
5. Bojarski, M.; Yeres, P.; Choromanska, A.; Choromanski, K.; Firner, B.; Jackel, L.; Muller, U. Explaining how a deep neural network trained with end-to-end learning steers a car. *arXiv* **2017**, arXiv:abs/1704.07911.
6. Bojarski, M.; Choromanska, A.; Choromanski, K.; Firner, B.; Jackel, L.; Muller, U.; Zieba, K. VisualBackProp: Efficient visualization of CNNs. *arXiv* **2017**, arXiv:1611.05418v3 [cs.CV].
7. AWS DeepRacer Documentation. Available online: https://docs.aws.amazon.com/deepracer/?icmpid=docs_homepage_ml (accessed on 3 October 2022).
8. Udacity's Autonomous Car. Available online: https://github.com/udacity/self-driving-car-sim (accessed on 7 November 2022).
9. CARLA Open-Source Simulator for Autonomous Driving Research. Available online: https://carla.org/ (accessed on 26 September 2022).
10. Autorope. Donkeycar: A Python Self Driving Library. Available online: https://github.com/autorope/donkeycar (accessed on 2 May 2022).
11. Cota, J.L.; Rodríguez, J.A.T.; Alonso, B.G.; Hurtado, C.V. Roadmap for development of skills in Artificial Intelligence by means of a Reinforcement Learning model using a DeepRacer autonomous vehicle. In Proceedings of the 2022 IEEE Global Engineering Education Conference (EDUCON), Tunis, Tunisia, 28–31 March 2022. [CrossRef]
12. Bojarski, M.; Chen, C.; Daw, J.; Değirmenci, A.; Deri, J.; Firner, B.; Flepp, B.; Gogri, S.; Hong, J.; Jackel, L.; et al. The NVIDIA PilotNet Experiments. *arXiv* **2020**, arXiv:2010.08776v1 [cs.CV].
13. Óscar, P.G.; Rafael, B.; Elena, L.G.; Luis, M.B.; Carlos, G.H.; Rodrigo, G.; Alejandro, D.D. Deep reinforcement learning based control for Autonomous Vehicles in CARLA. *Multimed. Tools Appl.* **2022**, *81*, 3553–3576.
14. Óscar, P.G.; Rafael, B.; Elena, L.G.; Luis, M.B.; Carlos, G.H.; Alejandro, D.D. Deep Reinforcement Learning based control algorithms: Training and validation using the ROS Framework in CARLA Simulator for Autonomous applications. In Proceedings of the 2021 IEEE Intelligent Vehicles Symposium (IV), Nagoya, Japan, 11–17 July 2021.
15. Terapaptommakol, W.; Phaoharuhansa, D.; Koowattanasuchat, P.; Rajruangrabin, J. Design of Obstacle Avoidance for Autonomous Vehicle Using Deep Q-Network and CARLA Simulator. *World Electr. Veh. J.* **2022**, *13*, 239. [CrossRef]
16. Gutiérrez-Moreno, R.; Barea, R.; López-Guillén, E.; Araluce, J.; Bergasa, L.M. Reinforcement Learning-Based Autonomous Driving at Intersections in CARLA Simulator. *Sensors* **2022**, *22*, 8373. [CrossRef] [PubMed]
17. Wang, C.Y.; Bochkovskiy, A.; Liao, H.Y.M. YOLOv7: Trainable bag-of-freebies sets new state-of-the-art for real-time object detectors. *arXiv* **2022**, arXiv:2207.02696.

MDPI
St. Alban-Anlage 66
4052 Basel
Switzerland
www.mdpi.com

Electronics Editorial Office
E-mail: electronics@mdpi.com
www.mdpi.com/journal/electronics